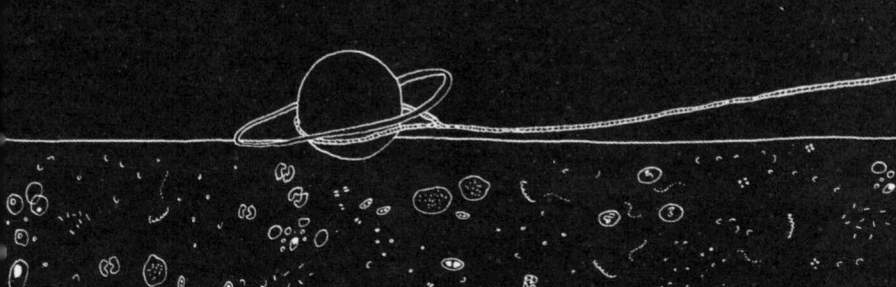

PENGUIN BOOKS

Eating the Sun

New York Times bestselling author Ella Frances Sanders is the writer and illustrator of two previous books, *Lost in Translation: An Illustrated Compendium of Untranslatable Words from Around the World* and *The Illustrated Book of Sayings: Curious Expressions from Around the World*, which have both been translated into eight languages. She currently lives and works in London, without a cat.

Eating the Sun

SMALL MUSINGS

ON A VAST UNIVERSE

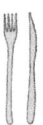

ELLA FRANCES SANDERS

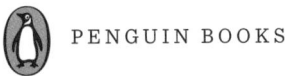 PENGUIN BOOKS

PENGUIN BOOKS

An imprint of Penguin Random House LLC

penguinrandomhouse.com

Illustrations by the author

LIBRARY OF CONGRESS CATALOGING-IN-PUBLICATION DATA

Names: Sanders, Ella Frances, author.

Title: Eating the sun : small musings on a vast universe / Ella Frances Sanders.

Description: New York, New York : Penguin Books, an imprint of Penguin Random House LLC, [2019]

Identifiers: LCCN 2018025207 (print) | LCCN 2018027516 (ebook) | ISBN 9780525504948 (ebook) | ISBN 9780143133162 (hardcover)

Subjects: LCSH: Astronomy—Juvenile literature. | Astronomy—Miscellanea. | Astrophysics—Juvenile literature. | Astrophysics—Miscellanea. | Universe—Juvenile literature. | Universe—Miscellanea.

Classification: LCC QB46 (ebook) | LCC QB46 .S2375 2019 (print) | DDC 520—dc23

LC record available at https://lccn.loc.gov/2018025207

Printed in the United States of America

1   3   5   7   9   10   8   6   4   2

Set in Sentinel Book

Designed by Sabrina Bowers

To the ones who
do not regard me
strangely.

INTRODUCTION

A sense of wonder can find you in many forms, sometimes loud-ly, sometimes as a whispering, sometimes even hiding inside other feelings—being in love, or unbalanced, or blue.

For me, it is looking at the night for so long that my eyes ache and I'm stuck seeing stars for hours afterwards, watching the way the ocean sways itself to sleep, or as the sky washes itself in colors for which I know I will never have the words—a world made from layers of rock and fossil and glittered imaginings that keeps tripping me up, demanding I pay attention to one leaf at a time, ensuring I can never pick up quite where I left off.

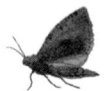

When one is considering the universe, unseen matter, our small backyard of the stuff, I think it is important, sensible even, to try and find some balance between laughter and uncontrollable weeping.

Cry because we cannot even begin to understand how beau-tiful it is, cry because we are terribly flawed as a species, cry be-cause it all seems so shockingly improbable that maybe our existence could be nothing but a dreamscape—celestial ele-phants in rooms without walls.

But then? Surely, we can laugh. Laugh because being riddled head-to-toe with human emotions while trying to come to terms with just how indisputably tiny we are in the grand scheme of things, makes absolutely everything and everyone seem quite ri-diculous, entirely farcical. We have heads? Ridiculous! There are arguments about who is in charge here? Ridiculous! The universe is expanding? Ridiculous! We feel it necessary to keep secrets? Ridiculous.

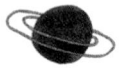

A lot of our time is spent trying to tie up loose ends, trying to shape disorder into something recognizably smooth, trying to escape the very limits that hold us close, happily ignoring rough edges and the inevitable. We separate ourselves out into past, present, and future, if only to show that we have changed, that we know better, that we have understood something inherent; if only to draw neat lines from start to finish without looking back.

The problem is that chaos is always only ever sitting just across the table, frequently glancing up from its newspaper, from its coffee cup filled with discolored and imploding stars. Because chaos too waits. Waits for you to notice it, for you to realize it's the most dazzling thing you've ever seen, for all of your atoms to collectively shriek in belated recognition and stare, mouth open, at how exquisitely embedded it is in everything. Because we are not designed to be more orderly than anything else; seams have a tendency to come apart with time—you and the universe are the same in this way, which makes for a delicately overwhelming struggle.

So, then, if you can't ever end things neatly, can't ever put them back quite the way you found them, surely the alternative is to remain stubbornly carbonated with possibility, to never rest from your rotation. To keep assembling stories between us, stories about how everything was everything, about how much we loved.

My hope is that this book is a small part of one such story.

"I break open stars and find nothing and again nothing, and then a word in a foreign language."

ELISABETH BORCHERS

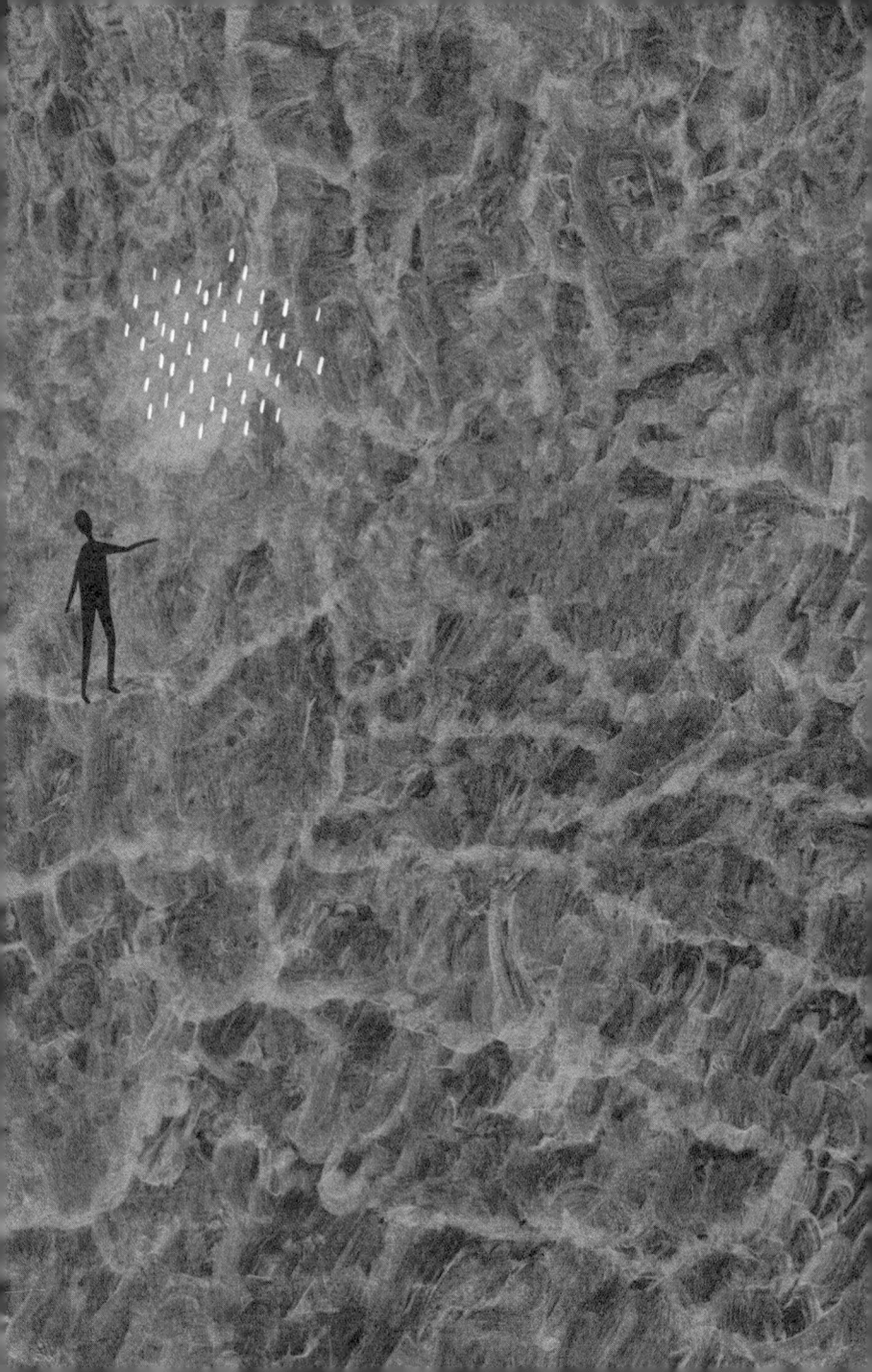

I AM MADE FROM CARBON

You are made from the remnants of stars.

Strung up like fairy lights, unobtrusive, at once quaint and overwhelming in a way only the impossible can be, the stars are to thank for your singular fragile body.

When stars die, they take the equivalent of one last deep breath and then fall in on themselves, like a soufflé that has been cooked for slightly too long. When this happens, they throw off their outer layers, releasing their contents to the magnificent nothingness and absolute everything that is the universe. Each year 40,000 tons of this starry dust falls to Earth; it contains the elements that will be used ceaselessly, throughout every living thing, around the entire planet.

Your body is composed of the products of such cosmic events, those remnants of burning giants. And oh, do they burn. Young stars, similar to the one you and I so fondly call "Sun," are mostly hydrogen, and with centers that measure 10 million degrees centigrade, their hydrogen becomes helium, and that helium slowly builds to form carbon, nitrogen, oxygen, iron, and every element that we walk around and into—everything that we are.

Depending on where you look, what you touch, you are changing all the time. The carbon inside you, accounting for about 18 percent of your being, could have existed in any number of creatures or natural disasters before finding you. That particular atom residing somewhere above your left eyebrow? It could well have been a smooth, riverbed pebble before deciding to call you home.

You see, you are not so soft after all; you are rock and wave and the peeling bark of trees, you are ladybirds and the smell of a garden after the rain. When you put your best foot forward, you are taking the north side of a mountain with you.

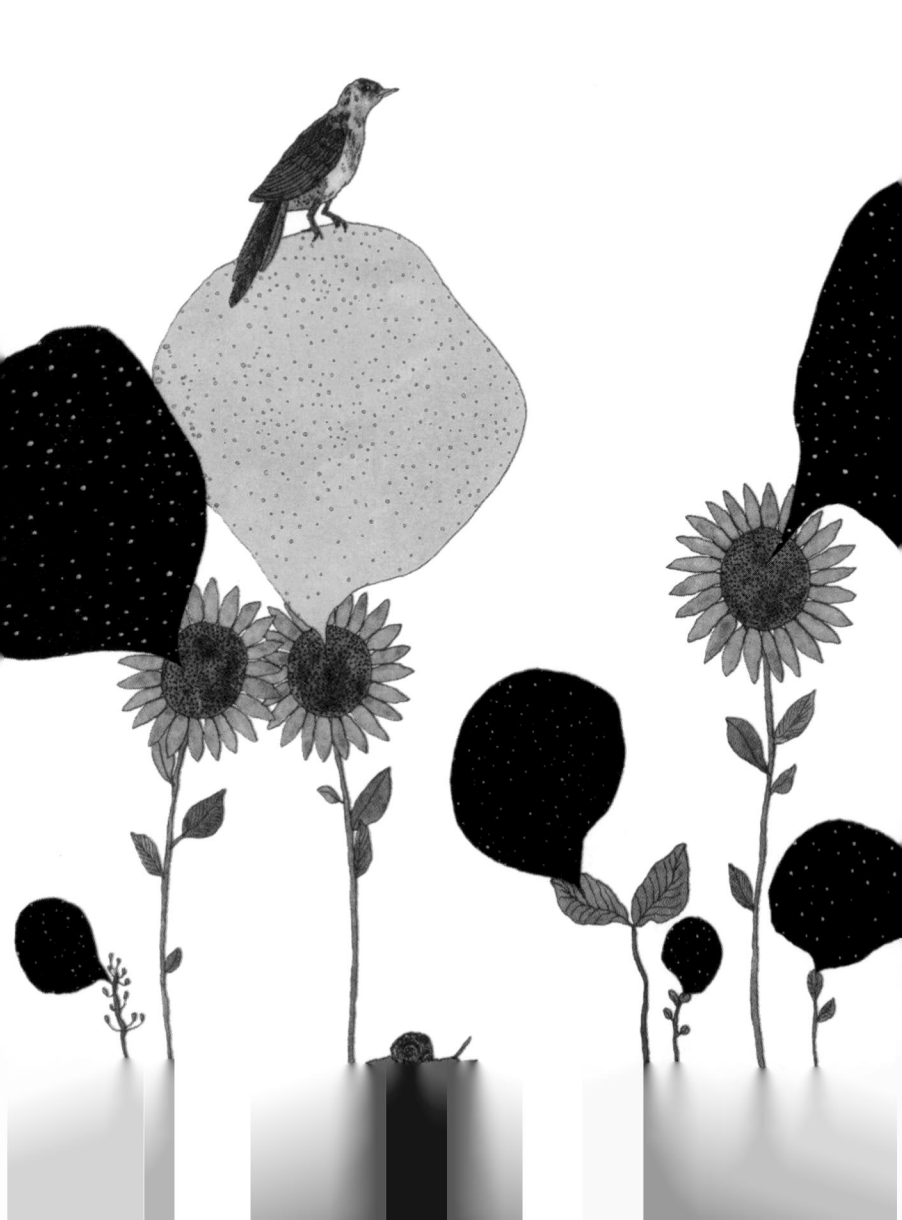

EATING THE SUN

You are what you eat, and we are all eating the sun.

While the sun is magnificent, having been happily burning for billions of years and likely to continue for billions more, we tend to only notice it once or twice a day, maybe more if we are driving into it or waiting for the washing hung out on a line to dry.

But today, if you ate a plant or an animal that ate a plant (or maybe you plan on eating half a grapefruit later), you've eaten the sun—light and energy and stories cut short.

Nearly all plant species carry out a process called photosynthesis, which allows them, with only water, chlorophyll, carbon dioxide, and some light energy from the sun, to create their own nutrients (which, in turn, become ours). During the first stages of this process, the energy of the sun splits existing water molecules into oxygen and hydrogen—the oxygen is given back into the atmosphere by the plant, and so we breathe. The hydrogen is used in the making of glucose, which the plant uses as energy to grow—to sway a little from side to side, track the passing of time, notice the inquisitive fingers of a careful few stroking their leaves. It is this digestible sun fuel that we are consuming.

Unlike plants, we animals cannot obtain our energy directly from fiery stars. In fact, we are wildly inefficient when it comes to moving and breathing and thinking about the person who went out of their way to smile at us at 3:22 p.m. yesterday, and all this leaves us entirely at the mercy of vegetation.

It's astonishing to think that we have been solar powered since the beginning of anything at all.

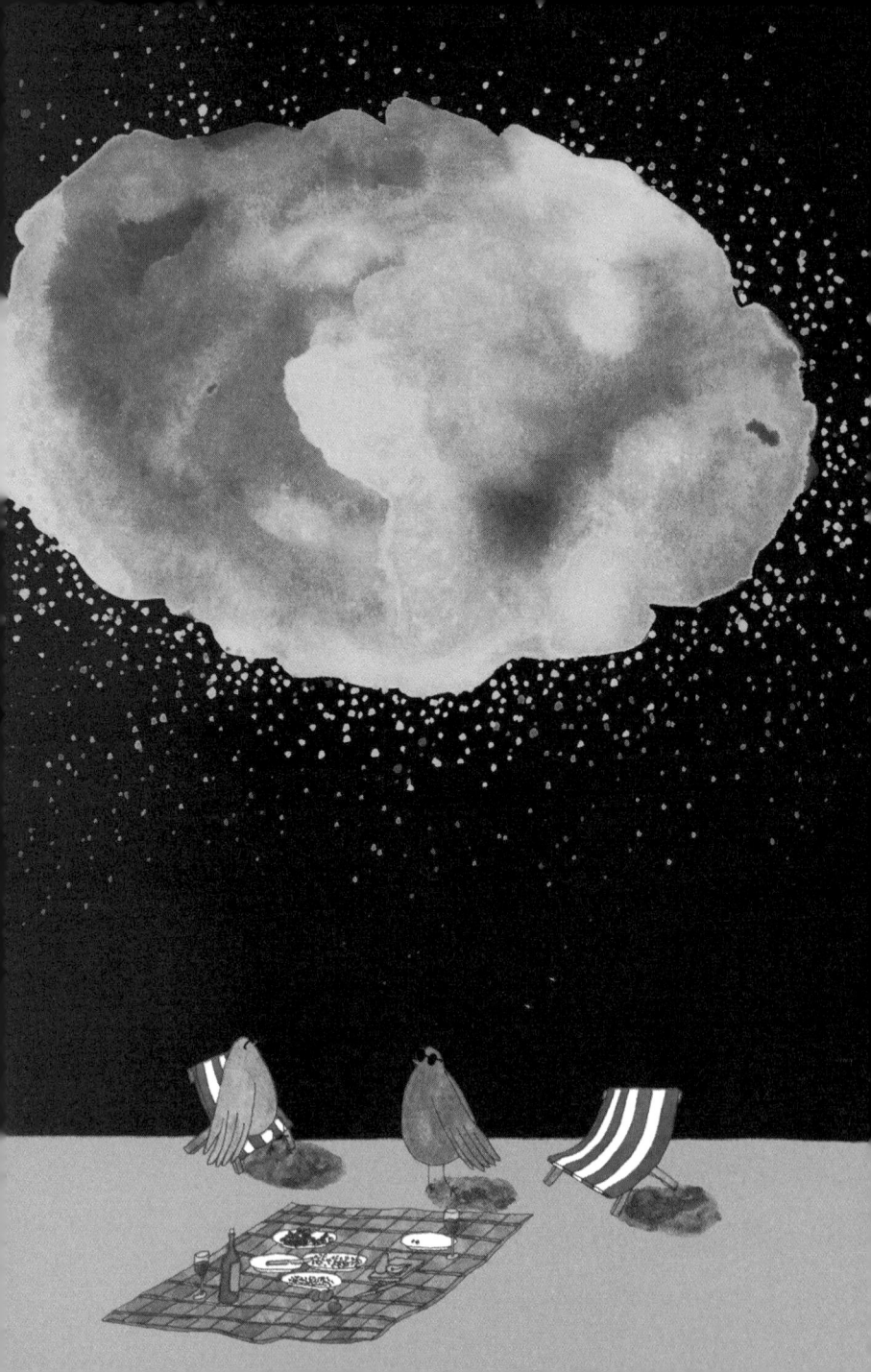

THE MOST LUMINOUS OBJECTS
IN THE KNOWN UNIVERSE

Your luminosity is intrinsic, but your brightness will depend on who is looking at you. In astronomy, the luminosity of an object is the total amount of energy that it emits across all wavelengths, measured over time. It is often used when referring to stars, whose luminosity will depend on their size, and mass, and temperature. Brightness (formally known as *apparent* brightness), although related to luminosity, varies wildly depending on the location, positioning, or proximity of the observer. Something with great luminosity might, to us, seem nothing more than a fleck of dust, only because it's sitting and burning and minding its own business unthinkably far away.

For a human standing on the surface of the planet, the brightest object is the nighttime star Sirius, primarily because it is a mere 8.6 light-years away. It is by no means the most luminous star, though, and even within the constellation containing Sirius, Canis Major, at least three other stars are thousands of times more luminous; they only appear fainter as they are so much farther away. Even the most ordinary of stars seem noteworthy from where we are, and so we point at those pins of ancient light, nod at the brightness, assign them names and neighbors.

In February 1963, a Dutch astronomer named Maarten Schmidt was analyzing an unusually bright speck in the sky, slowly realizing that while he had thought it might be a nearby star, it actually was something entirely different: not close at all, but rather 2 billion light-years away, and in order for it to still be so bright at that distance it would have to be more luminous than anything known at the time. Schmidt named this object a "quasar," which is short for "quasi-stellar object," or QSO. Named 3C 273, it is located in the constellation Virgo, and optically speaking is the brightest of the bunch.

In the fifty or so years since this discovery, hundreds of thousands of quasars have been observed. They remain some of the most astonishing things in the universe, and are perhaps the most luminous of all. Lying in the middle of galaxies, galaxies with vast black holes that can be billions of times larger than the sun, the temperature of a quasar can reach tens of millions of degrees, and their immense radiation means they outshine everything around them, drowning out all nearby stars. But they are not unchanging, and while one minute a quasar might be blinding, ten years later it can have become just another average galaxy. In astronomical terms, ten years is the briefest of moments, but it is events and observations such as this that lead to a better understanding of a black hole's appetite: how they can be ravenously hungry one moment and completely disinterested the next.

PLANETARY MOTION

Within the little, vast pocket of universe that is our solar system, the sun is the largest thing by far, about one thousand times heavier than the largest planet, Jupiter. It is around this burning star that we all circle, for much the same reasons that our moon orbits us—gravity, velocity, an apparent magic (see page 45).

Nearly everything in our solar system rotates on its axis in the same direction the sun does, a "prograde" rotation, but some things, like the planets Venus and Uranus, are oddities. Venus rotates in an opposite, or "retrograde," direction, completing one rotation every 243 Earth days. The rotation of Uranus is even more peculiar, as it's tilted rather dramatically on its side at a near-right angle and doesn't seem to know quite what it's doing. But for nearly everything else, their all-together-now spinning goes back to the beginning; the Milky Way galaxy that our solar system resides in was formed by a rotating mess of gas and dust, and as things keep moving unless given a very good reason to stop, we're all still spinning.

"Satellite" is a term that can refer either to something out there already, such as any moon or Earth (in relation to the sun), or to something man-made, such as the International Space Station. If it has a regular, repeated elliptical path for an orbit, then it's almost certainly a satellite. Although technically possible, orbits are almost never perfectly circular—however small, there will be some deviation from a perfect route. In the case of Earth's orbit around the sun, the deviation is only three degrees, which means if you were to draw a circle 100 meters across, we only move away from that perfection by 14 millimeters (diagrams or drawings of the solar system often depict misleadingly stretched elliptical paths).

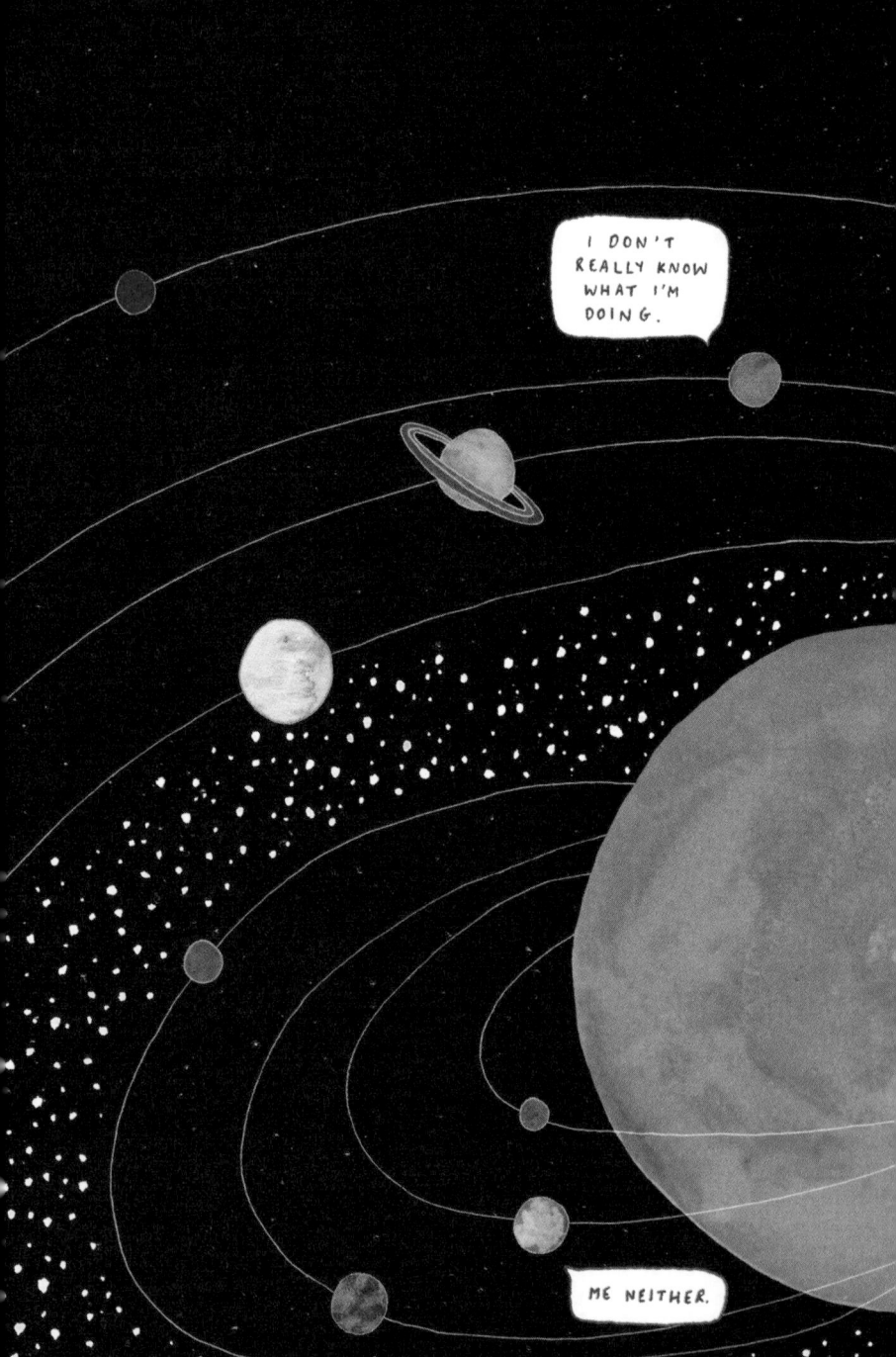

If anything, we should be glad that orbits are not perfectly circular, as it means we can use words to describe them that are enough to make a person weak at the knees, like "perihelion" (when a planet is at its closest point to the sun) and "aphelion" (its furthest). These slightly imperfect orbits exist primarily because although the pull of the sun is immense, it is not strong enough to keep things consistently close: the farther away from the sun a planet travels, the slower it becomes, until it reaches its aphelion, where it will start to "fall" back in and pick up speed, moving its fastest when closest to the sun.

While it appears as though smaller objects, such as Earth, circle around unmoving larger ones, such as the sun, everything is in fact busy orbiting around a combined center of mass, called the barycenter. This is often so close to the center of the largest object that it appears to be static, but the barycenter actually moves a little depending on where planets are in their paths. While every single mote of dust in our solar system is technically enraptured by this barycenter, it is not surprising that we often defer to the sun as being the center of everything, because were you to weigh the whole solar system, the sun would account for a grand 99.87 percent, and is therefore rather winning when it comes to the gravitational game.

Until you know precisely (or even vaguely) how and why the celestial bodies in our solar system work, their dancing is easy to take for granted. But once you do know, it is difficult to tear yourself away. Each of our unassuming neighbors, busy with a slow, dimly lit waltz during both long days and veiled nights—never stopping for breath, never hearing any applause, but certain that they must keep going.

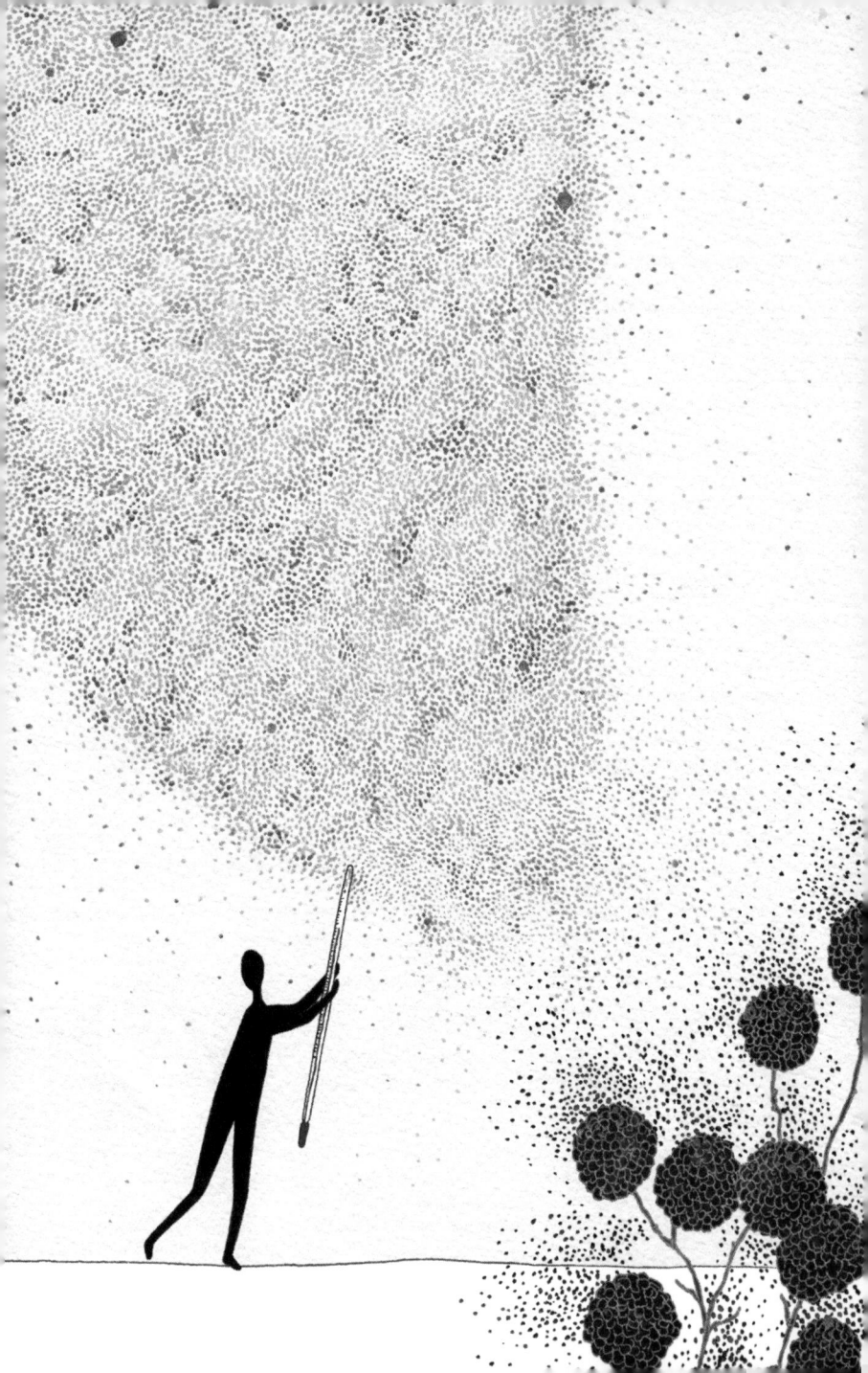

WHAT IS HEAT

"Heat" is really referring to thermal energy, an energy that results from the movement of particles—atoms, ions, and molecules—that form the gases, liquids, and solids of our perceived world. A window frame, an iceberg, your glass of water half empty or half full: absolutely everything contains heat energy.

Particles act in a way similar to people in large crowds, as they are constantly coming into contact with one another, jostling for room, an elbow here and a spectacular crash there. It is these continual collisions that form the basis of "kinetic theory," an idea developed in the late 19th century by a small group of suitably brilliant physicists.

If the temperature of something drops, there is a decrease in the kinetic energy of its particles. Conversely, with an increase in temperature comes an increase in kinetic energy. The warming and cooling of all things is really only a transfer of heat, and so a temperature is just a measurement of the ability of one thing to transfer heat to another—from me to you, coffee to spoon, here to there.

The reason heat moves from hot things to cold ones and never the other way around was explained concisely by an Austrian scientist called Ludwig Boltzmann, who published a series of papers in the 1870s. It is, strangely, only a question of probability, of sheer chance. Boltzmann realized that changes in temperature do not occur because of some absolute and unbendable law, but instead because it is statistically more likely that the fast-moving atoms of a hot substance will collide with the slower, cooler ones of a cool substance. When there are enough collisions, the thermal energy becomes more evenly distributed, and as this happens the temperatures of two things touching will begin to equalize, until a state called "thermal equilibrium" is reached, where the temperatures are the same, and the exchange of energy ceases.

Heat is decisively indecisive and generally won't stay where you put it—leave some in one place, turn your back for five minutes, and you'll find that it's gone someplace else entirely.

HOW TERRIBLY ILLUMINATING

We can think of light, in simple terms, as a way of transferring energy through space and throughout the universe. But when we typically throw the word "light" around as something to be admired and sat in and pored over, we are referring to only one part of a whole spectrum: the visible or optical part. However, that which our eyes can see is not alone.

Visible light sits in the middle of an electromagnetic spectrum. This spectrum includes radiation of all sorts—some with long wavelengths and low frequencies, like radio waves; others with short wavelengths and high frequencies, like X-rays. All of this radiation moves at the speed of light, 299,792,458 meters per second, in such a way that it cannot neatly be defined as either particles or waves. Physicists sidestep this small but insistent conflict of definition by calling it "wave-particle duality."

Light of any sort anywhere is created when excited atoms drop from a higher state of energy to a lower one, or move from a lower state to a higher one; they gain or lose energy when doing this, and this energy is emitted in the form of a photon. The general term for making light by "exciting" atoms in this way is "luminescence," which is more than anything just a very nice word (unlike "nice").

Light behaves predictably, which means that we can shape and manipulate it, using it for processes that appear slightly magic, as if they were the result of people pulling invisible strings. Whether from the sun or some other suitably glowing source, light reflects off the things surrounding us—people, buildings, a bird in flight—and allows us to see a shape, a story we can later recount in shocking Technicolor.

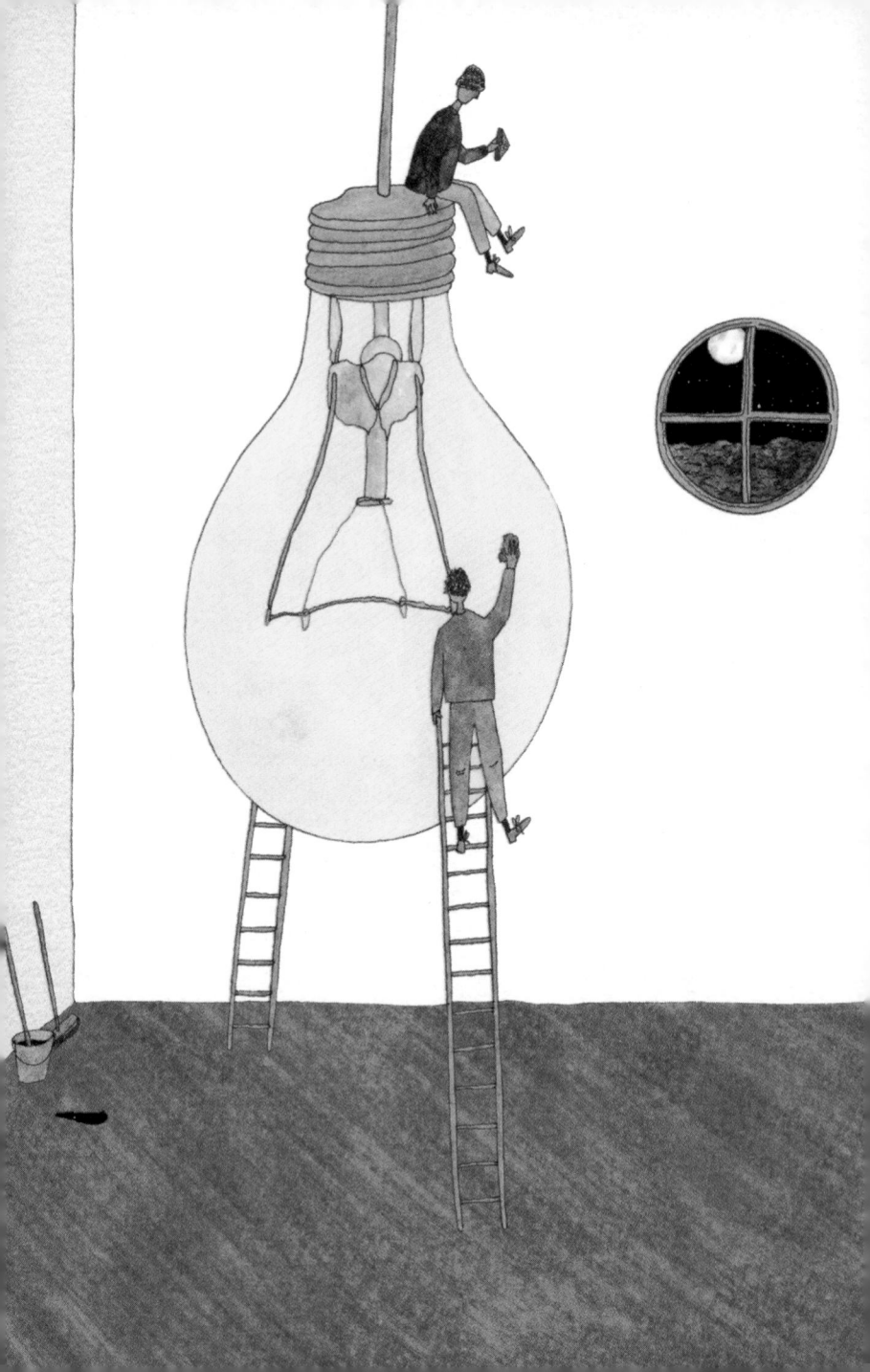

This particular behavior of light can be classified as either specular or diffuse reflection. Specular refers to light reflecting off something in a well-defined way, like a mirror, glass, or water smoother than stillness. Most reflection, though, is diffuse, because nearly all of life and everything in it is irregular and hard to predict—when light hits such things it scatters.

Light will also refract when it travels through certain things, and this "bending" of light is especially useful for those among us who need to wear spectacles in order to see decisions in the distance; those lenses we look through are refracting light.

Light, too, exhibits behaviors called diffraction and interference. Diffraction is the bending of light around obstacles, or the passing of it through gaps—when you half close your eyes in the dark looking at streetlamps, as they trick the light into bizarre lengths and nimble, clipped movements, you're seeing an example of diffraction. Interference is the meeting of two light waves—either they get along perfectly well and cancel each other out, or they disagree and together grow and change. The colored, unpredictable surface of a soap bubble is an example of interference, a phenomenon that may cause a person to think about washing the dishes in an entirely new way.

The sun, mighty thing that it is, radiates vast amounts of energy all the time, and while only a small portion of that reaches us on Earth, it's more than enough to light our days, even with an ever-present delay—the light you're seeing now still belonged to the sun eight minutes ago. But this cosmic lag doesn't make a sunset any less breathtaking, or any less timeless—you might be looking at a sun that's no longer there, but it still makes for good company.

The lack of a sun, and the consequent lack of light, would result in an existence not dissimilar to this one, but we would simply be unable to see any of it.

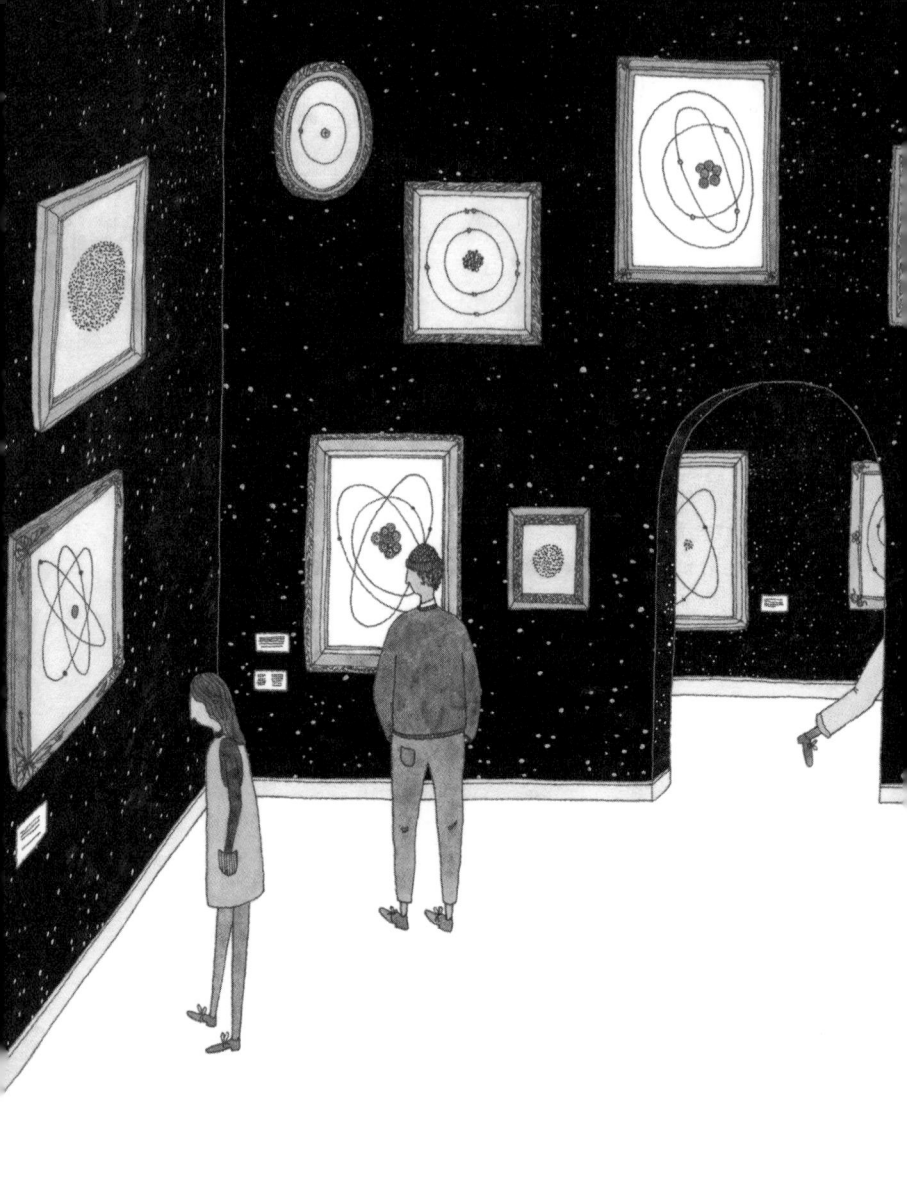

ATOMS ARE WORKS OF ART

Perhaps we should hang paintings of atoms in large, air-conditioned museums with white walls and stare, gaping at them in quiet amazement. Atom galleries, we can call them, and there we will point and say, "But look! How unthinkable that these tiny, unassuming things are responsible for it all."

During one of his lectures in the early 1960s, the abnormally brilliant American physicist Richard Feynman said this: "If, in some cataclysm, all of scientific knowledge were to be destroyed, and only one sentence passed on to the next generation of creatures, what statement would contain the most information in the fewest words? I believe it is the atomic hypothesis (or the atomic fact, or whatever you want to call it) that all things are made of atoms—little particles that move around in perpetual motion, attracting each other when they are a little distance apart, but repelling upon being squeezed into one another. In that one sentence, you will see, there is an enormous amount of information about the world, if just a little imagination and thinking are applied."

Atoms are absolutely crucial to our understanding of the universe, and although the belief that everything under and beyond the sun is made from them has been hanging around for over 2,500 years, it wasn't until the last two hundred that it became truly impossible to think anything else could be going on. Now we can not only see them, using laughably powerful microscopes, but we can even manipulate them; we can move that which makes up the entire universe, if only just a little.

Such a beautiful (and until recently invisible) idea, the importance and unavoidable nature of atoms, one that seems to put everyone and everything on a satisfyingly level playing field. Your good and bad decisions, your wingspan, your wholeness as a person—these are all possible because of your own 7 billion billion billion atoms, each one made up of (roughly speaking) a positive nucleus in the middle, and the negative electron cloud surrounding it—a cloud that sort of dances from side to side, alternately enchanting other atoms and pushing them away (the really complicated magic can be left to quantum mechanics). Without atoms, nothing would be here; not the book in your hands, not the pen that leaked into your pocket this morning, not those buildings that are enough to make you scared of heights, nothing. If it weren't for atoms, there wouldn't be mass, or molecules, or matter, or me, or you.

PLANTS BEHAVE BETTER

We humans are incredibly shortsighted compared to plants; while we come and go from this life with astonishing frequency, plants can live for hundreds, sometimes thousands of years, and this difference in time scale has probably in part led to our inability to protect and understand them in a collective sense.

But, phew, we are now trying to better understand them. While plants lack neurons, the type of nerve cell that transmits information and is commonplace in nearly all animals, they have their own kind of intelligence. The study of this is known as plant neurobiology, and scientists in the field have discovered that plants possess characteristics such as memory, learning, and problem-solving. However, while the word "intelligence" is used a lot when discussing these things, it must not be confused with consciousness, and neither of those with complexity. Speaking about them in anthropomorphic terms can be problematic, as plants have different priorities than we do—when we leave, they are always the ones who stay, dealing with whatever chaos we have left behind.

It is precisely because they have to remain fixed in one place that they have needed to evolve such nuanced and chemically complex ways to survive: a plant must have an extensive knowledge of its immediate surroundings because it needs to feed, grow, reproduce, and defend itself without ever going anywhere.

Trees make for dazzling examples of the complexities and intelligence of plants. Their root systems intertwine, both directly and through something called "mycorrhizal fungi," which live in their root systems and form one half of a supremely important symbiotic relationship. Without the fungi, trees wouldn't have enough access to minerals in the soil, and as the fungi lack chlo-

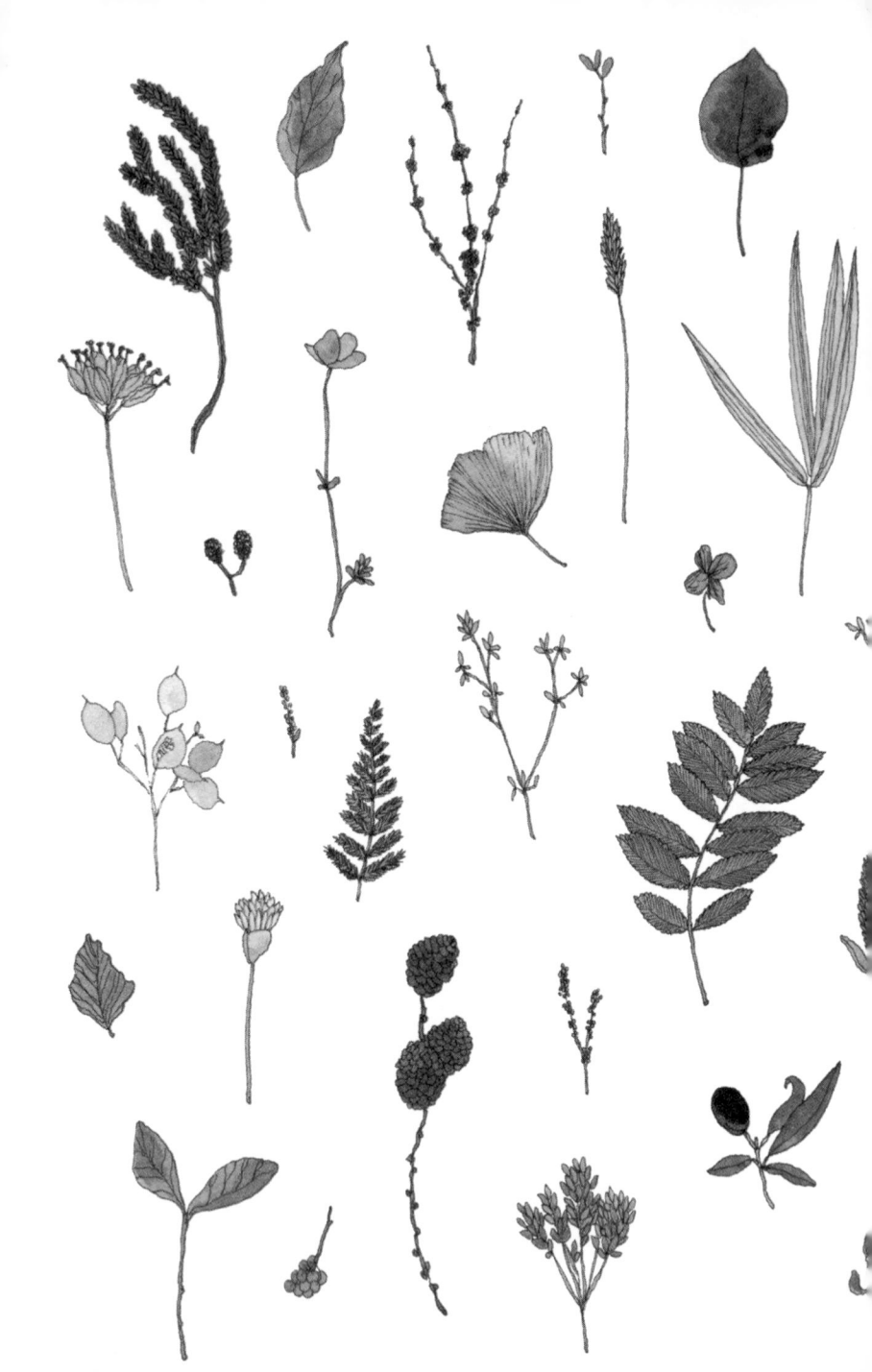

rophyll, they wouldn't be able to grow anywhere at all without the trees.

Trees are also seemingly able to distinguish their own roots from those of other species, and even those of their relatives. They share food and help to nourish their competitors when they are sick or struggling (in winter an aspen will likely not do as well as a conifer, so the conifer lends a hand), and all this apparently for no other reason than that living becomes much easier when you're helping others, rather than simply ensuring your own survival. In fact, trees' roots can sometimes end up so connected that two of them will die at the same time.

In a way, trees know exactly what they are doing, they just move far more slowly than we do. While the nerve impulses of a human can travel at speeds of 119 meters per second for muscle positioning, and 0.61 meters per second for pain signals, the electrical impulses of a tree move at only 0.00014 meters per second, or one-third of an inch per minute. They appear to us sluggish, but they are actually remarkably good at adapting in the face of stressors like climate, viruses, or changes in soil.

Although thousands of new species of plants are still being discovered every year, of the ones we do already know about, more than one in five is threatened with extinction. It seems that plants are so fundamental to humankind that we're prone to forgetting quite how much we rely on them for everything, from food, fuel, medicines, and materials, to their role in regulating the planet's temperature, climate, and composition. We expect and want too much, and it's showing up in increasingly obvious ways. In the words of the late ethnobotanist Tim Plowman, "They can eat light, isn't that enough?"

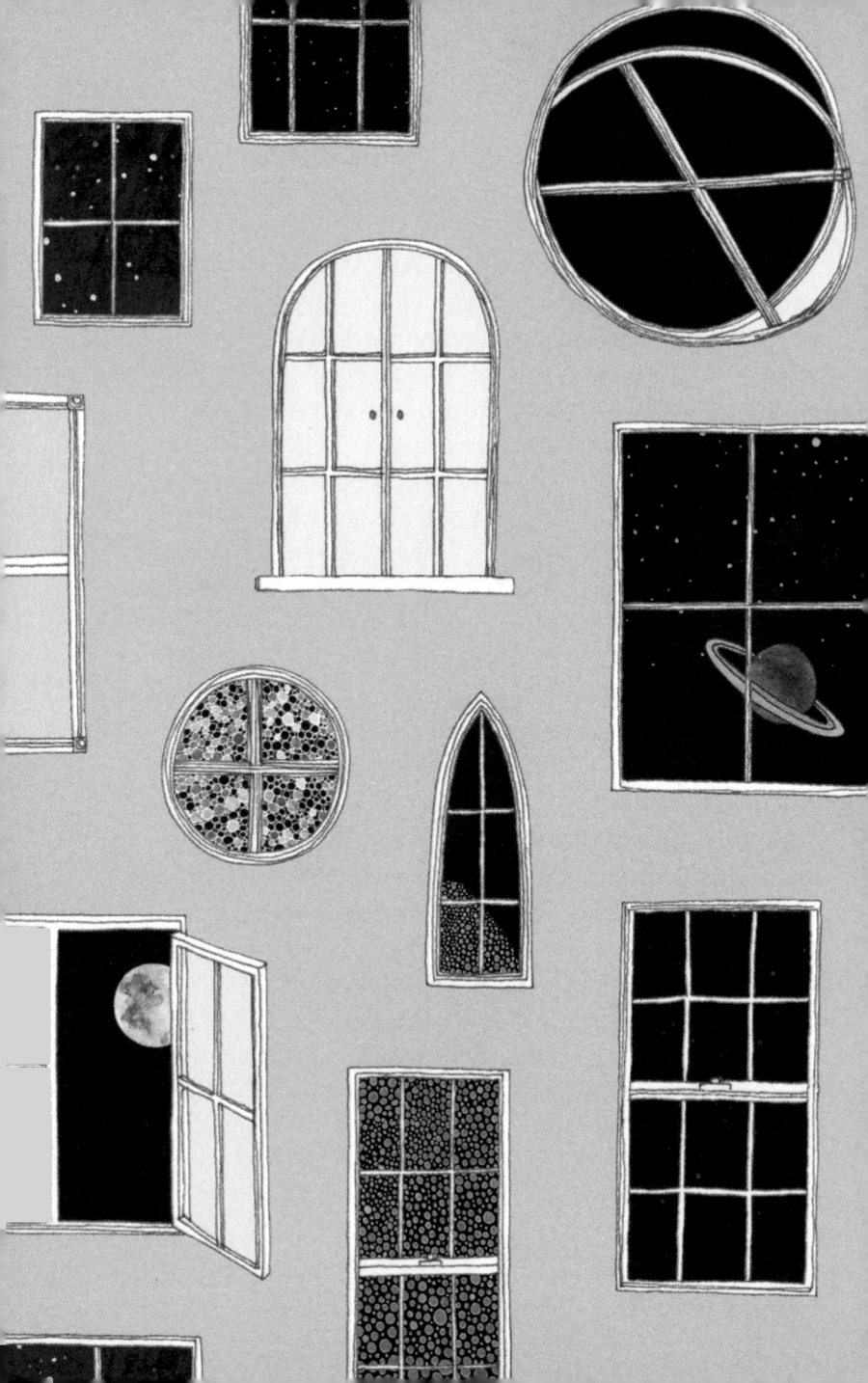

MILKY SOLAR GALAXY SYSTEMS

It's worth clearing this up.

Solar systems are commonplace, and ours contains a single star, which we call Sun, and this is surrounded by the planets, and by everything else that orbits either directly or indirectly (though never accidentally, because there are no accidents in space). These other objects include things like moons, asteroids, rocks, and a great deal of dust.

Our solar system sits within a galaxy called the Milky Way—a galaxy being an extremely large system of stars, isolated from other similar systems by vast regions of space. In the Milky Way, which is about 110,000 light-years across, our sun is just one of between 100 and 400 billion stars, most hosting their own planets. Galaxies, like egos, come in varying sizes, and while ours is fairly spacious, there are others like the Andromeda galaxy (our neighbor) that are much, much larger.

The term "universe" refers to all of the galaxies that have ever been and will ever be observed—the totality of all known and unknown objects and wonders throughout the cosmos. As late as the 1920s, astronomers thought that the Milky Way galaxy contained all of the stars within the universe, a thought which now seems almost endearing, as there are actually countless billions of them.

The thing is, we can only ever see so far. While advanced telescopes can reach and study billions of far-flung galaxies and phenomena and study them in exquisite detail, there are things that we will never, ever be able to hold onto—one consequence of living in a forever expanding universe.

In September 2014, astronomers realized that the group of galaxies the Milky Way is part of was one hundred times larger in mass and volume than they had previously thought: 500 million light-years from side to side. Galaxies have a tendency to nestle together in groups or clusters; the larger, more densely packed ones are known as "superclusters." As it turns out, we are living within one such supercluster, and having mapped the region, the astronomers named it Laniakea—Hawaiian for "immeasurable heaven."

If that doesn't make you feel lit up on the inside, I don't know what will.

YOU ARE NOT YOURSELF TODAY

The idea of an unchanging "you" or "self" is inherently fraught with confusion and conflict, and if you consider the topic for too long it can begin to feel clammy, almost suspect. An apparent string running through all the previous versions of you—the one five minutes ago, a few hours ago, several years—the idea of "self" inevitably gets tangled up in things like the physical body and appearance, like memory. It's clear that you cannot pin yourself down as any one particular "thing" but rather that you resemble a story line, an endless progression, variations on a theme, something that enables you to relate your present "self" to the past and future ones.

We do seem to make sense of ourselves and the world as a part of a narrative—we think in terms of main characters, those we speak and interact with, and where the beginnings, the middles, and the endings are. The brain is determined to make up stories and plotlines even when faced with compelling and contradictory evidence. And just about everything in our lives does have something to do with other people, or how they perceive us to be. We all think that we will or won't behave and act in certain ways, but we are often mistaken, surprised—we do things that are decidedly "out of character." Though frustrating, we cannot ever choose or control the aspects of life that ultimately influence what we say, do, or think.

Depending on where you are standing, there are many different and often conflicting ways to dwell on the question of self. The 18th-century Scottish philosopher David Hume said that the self was nothing more than a bundle of perceptions, and American philosopher Daniel Dennett describes it as a "center of narrative

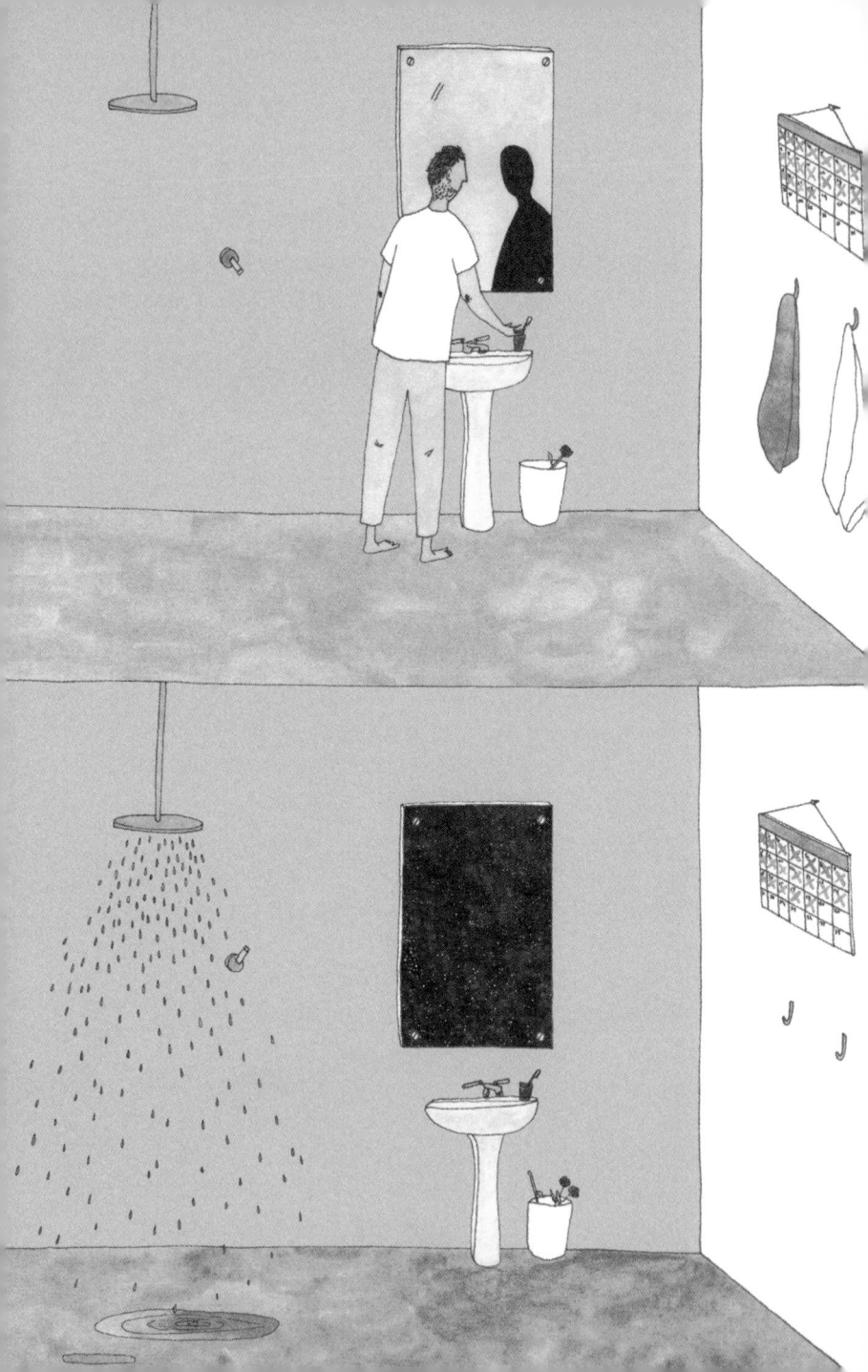

gravity." Meanwhile, social psychologist Hazel Rose Markus says that "you cannot be a self by yourself."

While you're probably not the shiny and singularly important entity you might believe yourself to be, we do seem to need some sense of self to get by, and it can be integral to many things in our lives—love, learning, what we pay attention to. But you are endlessly creating this "self," and it is not something that is just standing around waiting for you to find it. No matter how you want to consider this, however stuck you become in the questions that are inevitably kicked up, it can be perhaps faintly reassuring to remember that me, myself, and I, we contain multitudes.

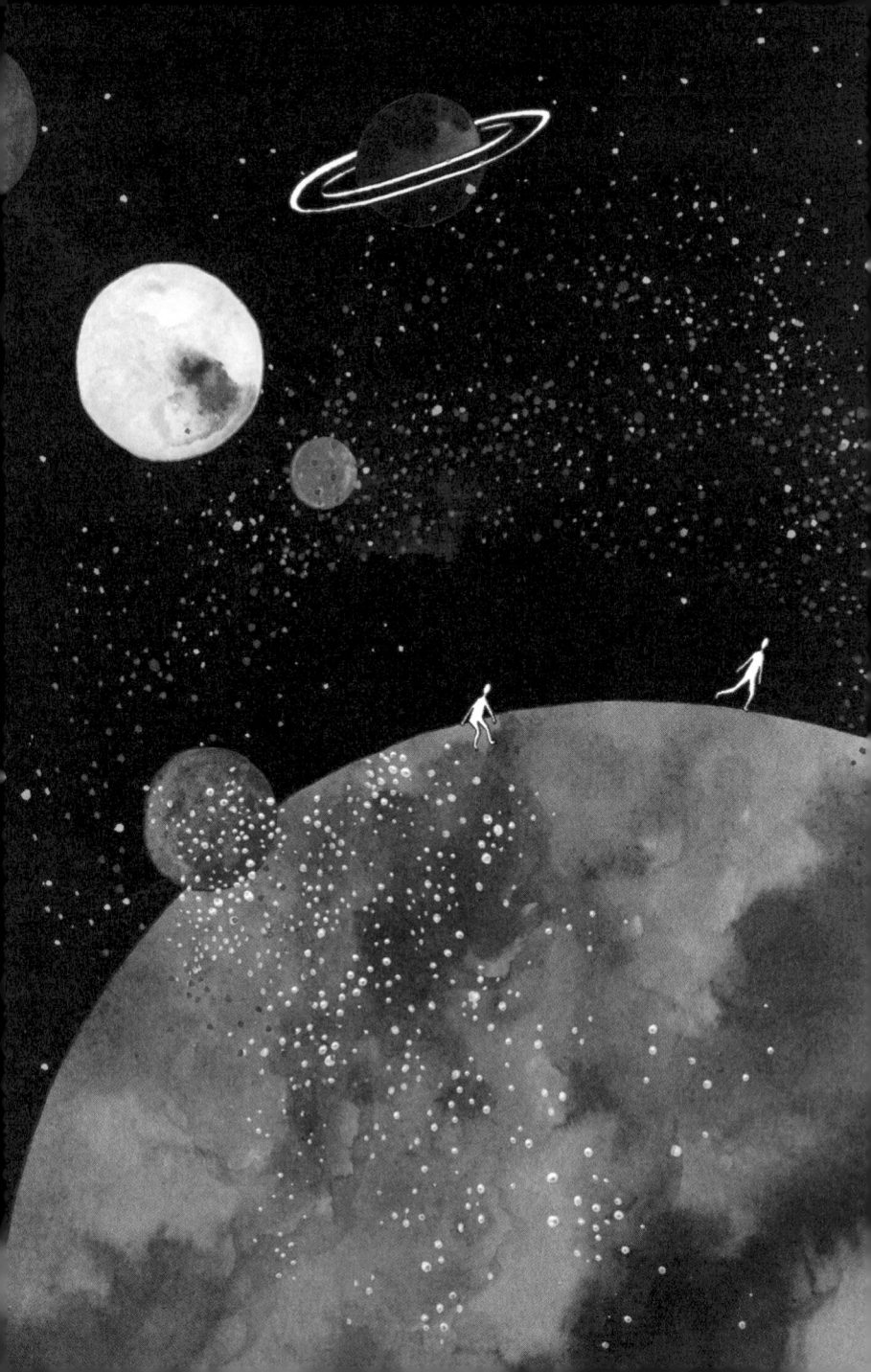

MITOCHONDRIAL EVE

We often find ourselves with a very individual hunger to know where we come from, who we are. Of course, the people who we think we are? They might not be us at all.

Genetically speaking, we humans are all very much the same, and we share 99.9 percent of our DNA (deoxyribonucleic acid) with any other person, anywhere. While that may sound astonishing, consider this: our DNA only differs by 1.3 percent from that of a chimpanzee. We also share 90 percent of our genes with cats, 80 percent with cows, about 60 percent with both chickens and fruit flies, even 50 percent with a banana. This seems both impressive and meaningless, as everything on Earth has a certain percentage of genetic overlap.

"Genome" is the term for a complete set of DNA, and your own genome contains instructions for developing and shaping all you ever were, all you are now, and all you ever will be. Every molecule of DNA consists of two twisting strands—which look to be stuck in the kind of awkward dance you get when both parties are afraid of getting inappropriately close—and each strand is made up of base pairs. The human genome contains about 3 billion of these base pairs, although only 0.1 percent of them differ between each of us. However peculiar it may sound, this seemingly minute variation is enough to account for all the differences in all the humans who have ever resided on the planet.

When looking backwards in genetics, we can reach what is called a "most recent common ancestor," or an MRCA—the most recent individual of a species from which all other organisms are directly descended. When only looking at humans in Europe, this common ancestor lived between 400 and 600 years ago, while the MRCA for all humans currently living on Earth is thought to have lived around 3,000 years ago.

It is possible to go back further, though, and roughly identify both a matrilineal (traced using the female line) and patrilineal (traced using the male line) common ancestor. The female ancestor is known as "mitochondrial Eve," and has been traced from mitochondrial DNA. This type of DNA is derived almost exclusively from the mother's side, and remains quite untouched between generations. Mitochondrial Eve is believed to have lived about 200,000 years ago. The male ancestor is known as "Y-chromosomal Adam," and has been traced through the Y chromosome, which is passed down between male individuals without any recombination. Y-chromosomal Adam is believed to have lived between 237,000 and 581,000 years ago.

What all this means for genealogists isn't yet clear, and there are still contrasting theories about how humans spread themselves about the planet—whether mitochondrial Eve and her descendants moved out of Africa and displaced all other types of man (known as the "Out of Africa" theory), or whether we populated the globe in a much more parallel, multiregional way. It is a struggle, as the mapping of humans using things like mitochondrial DNA is incomplete, and the fossil dating methods also used to establish timelines for humans can be hit and miss. It is likely summed up best by genetic anthropologist John Hawks, who said that "a large-scale reorganization of the science of human origins is upon us."

Our imperceptible differences are perhaps quite revelatory, but our sameness even more so—your DNA is entirely indifferent to what is written inside your passport; it is only concerned with a slow and ordered biological progression. The way we still insist on territories and borders, our frantic justification of cultural divides, it all begins to look strange and outdated and decidedly uncivilized in this genetic light.

It is clear that we haven't quite figured out how to be just yet.

I'LL BE WHERE THE BLUE IS

You would be hard-pressed to find someone who doesn't care for the color blue, and surely life would be quite unbearable for such a person—71 percent of Earth's surface is covered in glittering saline water, and the sky is often a persistent shade of cerulean. Given that the color is so all-encompassing, it seems strange to know that for much of human history (and with the exception of the Egyptians), we didn't even have a word for it.

Blue as a color in nature is actually incredibly elusive—while plants that are rich in anthocyanins are genuinely blue (a blueberry being perhaps the best example of this), the vast majority of animals do not produce this pigment, and so in most instances a "blue" creature only appears to be so, and it is in fact the result of iridescence, or selective reflection (more about light on page 13).

A phenomenon known as "Rayleigh scattering" is responsible for both the blue of the North American bird *Cyanocitta cristata*, or blue jay, and that of the sky. In the case of the jay, its feathers contain melanin and would appear black if it weren't for tiny air sacs in the feathers that scatter light, and so to our eyes the bird then seems rendered in endless variations of blue. And when we gaze skywards, we are observing sunlight entering Earth's atmosphere and colliding with particles in the air—compared to the other colors contained within light, blue has a shorter, smaller wavelength, and is therefore scattered more. The result of this scattering? Blue skies.

The ocean looks blue for much the same reason—water molecules absorb red, yellow, and green wavelengths of light, leaving the blue behind—but it can stray towards green or even red as light bounces off particles or sediment in the water. More than light, though, the largest impact on ocean color comes from

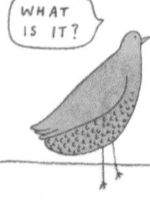
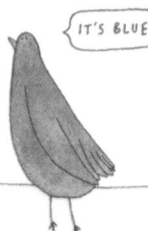

phytoplankton, tiny organisms which not only contain chloro-phyll and carry out photosynthesis, but also reflect the green light in the spectrum, meaning the water they thrive in appears much, much greener than oceans where they are scarce, like the clear waters of the Caribbean.

Prussian blue, dark blue, cobalt blue. Blink and you'll miss it blue.

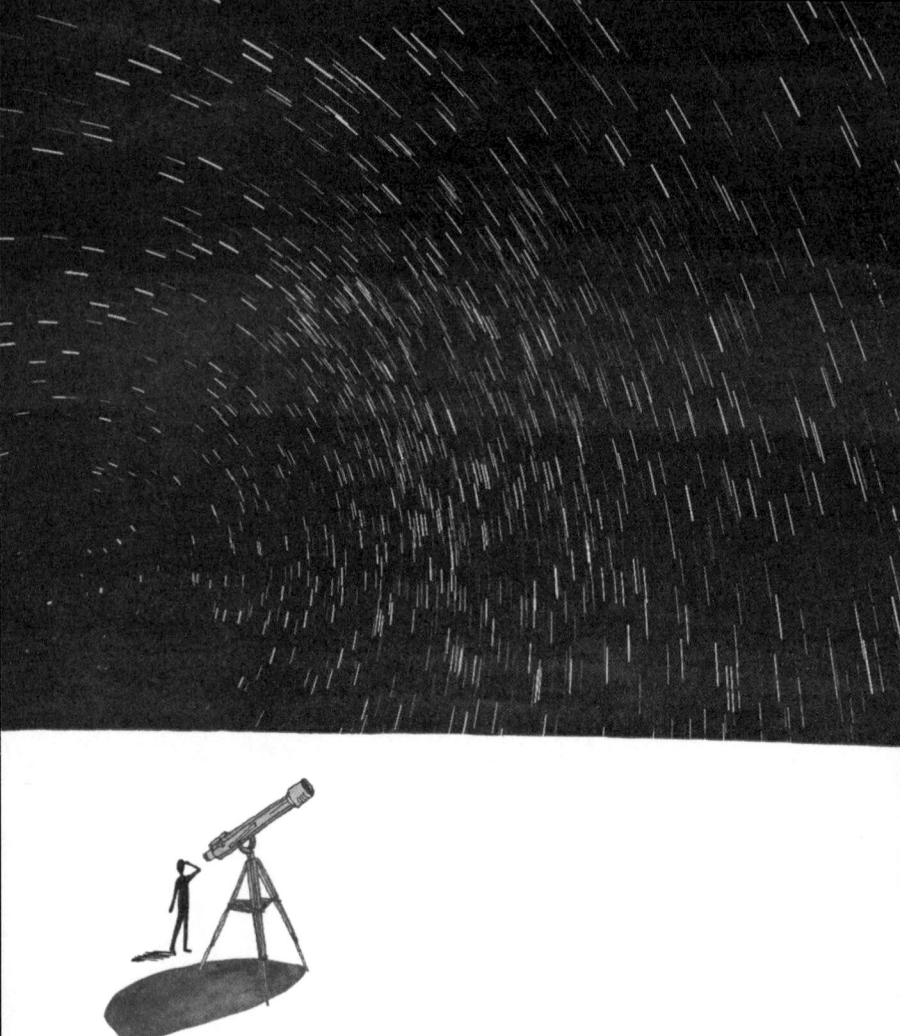

LONG - DISTANCE RELATIONSHIPS

The distances between stars can be hard to grasp. Astronomers measure such cosmically huge intervals in various ways, using things like the expansion of the universe, the color of stars (the apparent color relates to their surface temperature and age), and their varying degrees of brightness to determine distance. However, astronomers most commonly use something called parallax.

Parallax is simply the perceived difference in the position of an object: when something seems to have changed its location only because you have changed yours, like when you put a pencil in front of your face and look at it using only one eye at a time, and the pencil seems to have moved.

The particular movement and change in position of stars when compared to their background and distant objects is known as "stellar parallax." It is one of the oldest ways to measure distance in space; astronomers take note of a star, then do so again six months later in order to measure how far away the star is (the method involves a lot of triangles).

The term "parallax" forms the "par" in parsec, a unit of measurement the equivalent of 3.26 light-years, or 31 trillion kilometers (19 trillion miles). It is one of our best ways to measure the distances in space, but because of the great distances involved, even parsecs need to be used in multiples when expressing distances outside of the Milky Way. For things that are relatively close to our galaxy, you might use a kiloparsec; when talking about most other galaxies, megaparsecs; and for the really distant galaxies and most quasars (more on page 5), gigaparsecs (1 billion parsecs). The very edge of the known universe—something known as the "particle horizon"—lies more than 14 gigaparsecs, or more than 45 billion light-years, away.

Telescopes on the surface of Earth are limited in their accuracy to seeing stars up to about 100 parsecs away, as Earth's atmosphere affects the sharpness of a star's appearance. Thank goodness space-based telescopes do not suffer from such a limitation; they can better measure distances to and between objects far beyond our own stuck-on-solid-ground observations.

Accurate or "direct" distances are based on something called the astronomical unit, or AU, which is the distance between Earth and our sun, and are only possible to know when dealing with objects less than 1,000 parsecs away from us. All other distances are calculated using the connections between the different methods, as they all relate to and affect each other—collectively, these methods are known as the "cosmic distance ladder," or "extragalactic distance scale."

In short, it makes a person feel precisely, measurably, and overwhelmingly small.

CLOUDS TO BREAK YOUR HEART

At any given time, clouds are likely to be covering two-thirds of our planet, and being as intrinsic as they are to Earth's rhythms, it's not all that surprising that they decide many things on your behalf—your shoes, your willingness to wait, which method of transport you use. They are more than happy to make those decisions for you; perhaps this darkened sky will keep you indoors for days, perhaps that predominant blue will mean that you finally get around to planting those tulip bulbs. They talk, and you listen without realizing.

Clouds are difficult for meteorologists to model and predict, and yet are unfailingly present, reassuringly around. They are formed by water vapor or ice crystals hugging determinedly onto microscopic particles in the atmosphere known as "condensation nuclei"—things like smoke and dust and salt—and they do so because the air is simply too saturated to hold onto all that water anymore. When the water vapor condenses around such nuclei, cloud droplets are created—put millions upon millions of these together and you have a cloud perhaps weighing, although hard to believe, the equivalent of one hundred large elephants.

Although impossibly heavy, the water in a cloud is incredibly spread out, for miles, and the effect of gravity on something as minuscule as a single water droplet is barely felt at all. It is only when a droplet reaches a certain size that it weighs enough to fall as rain towards the two of us. Otherwise, they stay up there, and we paint them. We need their reflection and dispersing of solar radiation, their trapping of Earth's heat and their spitting of that back out again, their opposite cooling effect. The particles in a cloud treat all the different wavelengths of light equally, and this consistent scattering of the sun often gives them their classic

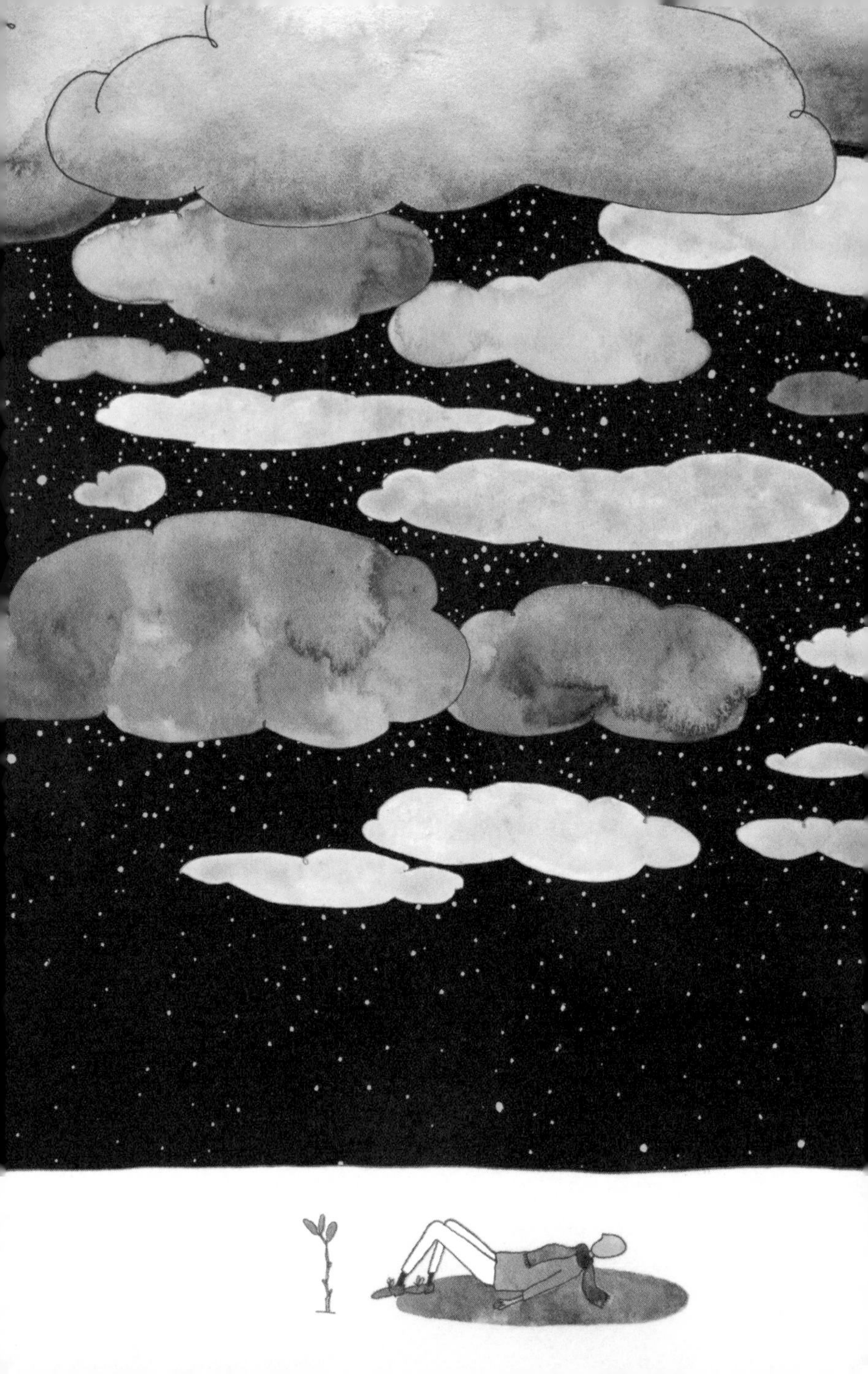

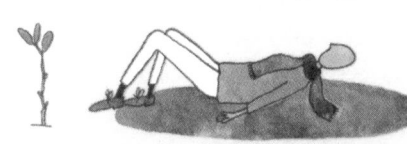

white appearance. Then things such as the thickness of a cloud or the position of the sun in the sky (for example, its lowness at sunrise) will cause differences within that overcast theme, providing clouds with their lyrical variations in color, those endless subtleties of white and gray.

From their poetic Latin names of classification—nimbostratus, noctilucent, cirriform, lenticularis—to their aesthetically pleasing yet ephemeral physical forms and their potential for destruction—summer rainstorms and hurricanes, the monsoon season and tornadoes out at sea—once you begin to notice how diverse and nuanced clouds are, they will inevitably break your heart (and break it in a way that hurts your neck).

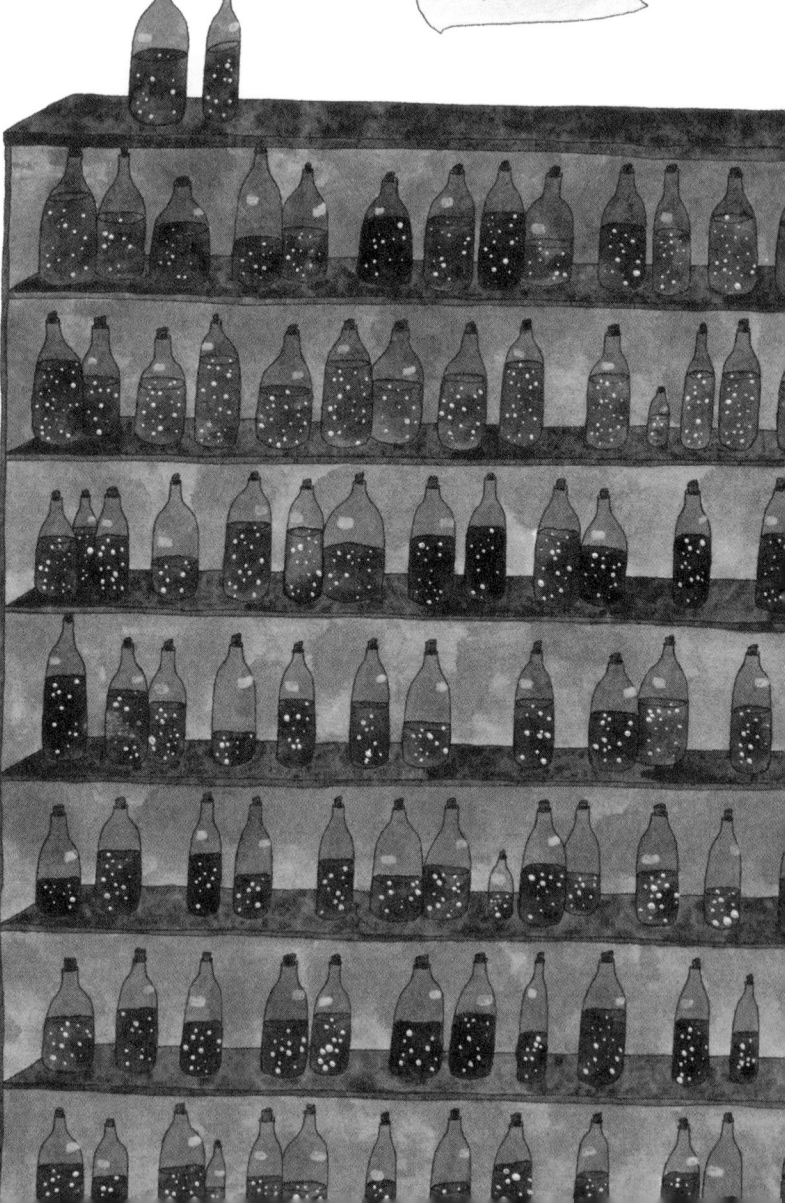

TIME

INDIVIDUALLY
PRICED

DOES ANYBODY ACTUALLY

KNOW WHAT TIME IS

Even though our concept and understanding of it continues to change and evolve, time remains one of the most difficult properties in our universe to pin down, as it can be alternately relative, imaginary, or real. On some fundamental level, it doesn't exist, at least not in a way you or I might recognize, and from one day to the next, much of what we refer to as "time" is either just memory or an anticipated future.

While ancient civilizations measured the passing of time with things like the annual flooding of the river Nile or the varying length of shadows cast by sundials, our modern understanding is built on Einstein's general theory of relativity, in which time is just a coordinate—it doesn't always run at the same speed, it's not just a simple line, and it exists within a space-time field of four dimensions.

Time is not symmetrical, but rather it has an inherent asymmetry, and a one-way direction. In 1927, the term and concept of an "arrow of time" was developed by a British astronomer called Arthur Eddington, who had realized that if time were to be symmetrical, the world would be rendered quite nonsensical. Such nonsense might not be immediately evident. For example, if a video of the planets orbiting the sun were played in reverse, you wouldn't be able to tell the difference between that and the same video being played forwards, and everything would appear to be in keeping with the laws of physics. But if a video of someone dropping a book to the floor were played backwards, it would look like the book is falling up—an absurdity. We remember the past, but we cannot remember the future.

The term "arrow of time" mostly refers to the thermo-dynamic arrow of time, which is neatly tied to the second law of thermodynamics, one of four laws discovered during the 19th century that define relationships between heat, work, energy, and temperature. This law states that entropy can only increase in a closed system—the closed system being our universe, entropy being a measure of disorder, mess. As time passes, so the entropy increases, and though we cannot measure time with entropy, we know that the energy in the universe moves slowly but surely to-wards an ultimate disorder. Things cannot be made any neater, and we cannot go backwards to yesterday—the second law of thermodynamics so casually imposes a direction on time.

There are other arrows of time, which vary in their connect-edness to each other, including a cosmological one, which points in the direction of the universe's expansion; a radiative arrow of time, involving the expanding outwards of waves from their source; a causal arrow, which has to do with cause preceding ef-fect; and a quantum one, which speaks of the symmetry of time and ties up with the famous Schrödinger equation, though no-body quite knows how this arrow relates to the other ones. There's also a psychological arrow: our perceived movement from a known past to an unknown future.

Culturally, the organization of time can be quite different, and this directly affects our experience of it. In some languages, the past is referred to as behind, and the future ahead, but in oth-ers, the past is ahead and the future behind, perhaps because the past can be seen, and in order to observe something, it needs to be in front of you, not behind. While some languages refer to time as a distance traveled, others refer to it as a growing volume—a long day, a full day. In English we think of it in linear terms, from left to right, but Chinese speakers think of time in terms of over and under, and in Greek time can be large, small. So easily do we mistake a word for the thing or phenomenon it speaks of, that it represents.

But it's alright, because when we gaze slightly bewildered at the night sky we are looking straight into the past; light may be traveling at 299,792,458 meters a second, but the distances mean that it doesn't arrive until we're ready to feel nostalgic. The top of your body ages ever so fractionally faster than your feet, because as gravity increases, time slows; the lined mountain ages faster than the ocean floor. And whether you are down or left, to the northeast or behind me, whether you call the day after tomorrow or I've just forgotten that you called at all, I'm certain that when we agree to meet in between order and chaos, you will be on time.

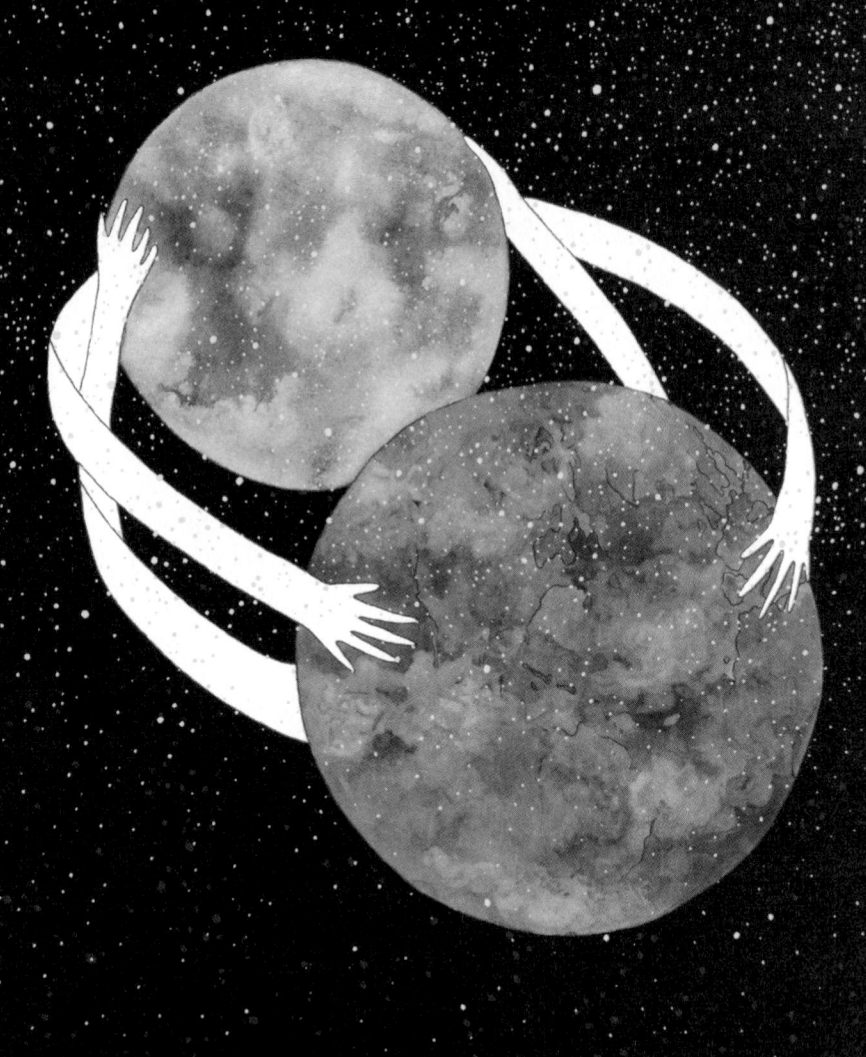

WHAT KEEPS THE MOON UP THERE

In our solar system, there is a lot of mutual attraction. Call it good timing, call it gravity, call it a love story that will never quite get to happen (it's mostly gravity).

As far as the Moon is concerned, it only has eyes for Earth. Earth is pulling on the Moon as it orbits us, exerting a "centripetal" or "center-seeking" gravitational force upon it, which is balanced perfectly by the Moon pulling on us, with a "centrifugal" or "center-fleeing" force, which tugs in the opposite direction. The Moon, lost in a world of its own, travels at 3,683 kilometers per hour, and unless interfered with, things in space like to keep moving just the way they are thank you very much (this is known as inertia).

It is this playing between the laws of physics—the velocity of the Moon, the equal pulling this way and that—that ensures the Moon never leaves our side (although it is getting farther away at a rate of 3.8 centimeters each year). These seemingly magical forces are present because anything with mass (a planet, a star) creates a curve in space-time around itself, and as the Earth is larger, the curvature in space-time it creates is large enough to affect the Moon, and dominate in such a way that it effectively "tells" the Moon to orbit. I guess the Moon doesn't mind, though, because it cannot ever bear to look away.

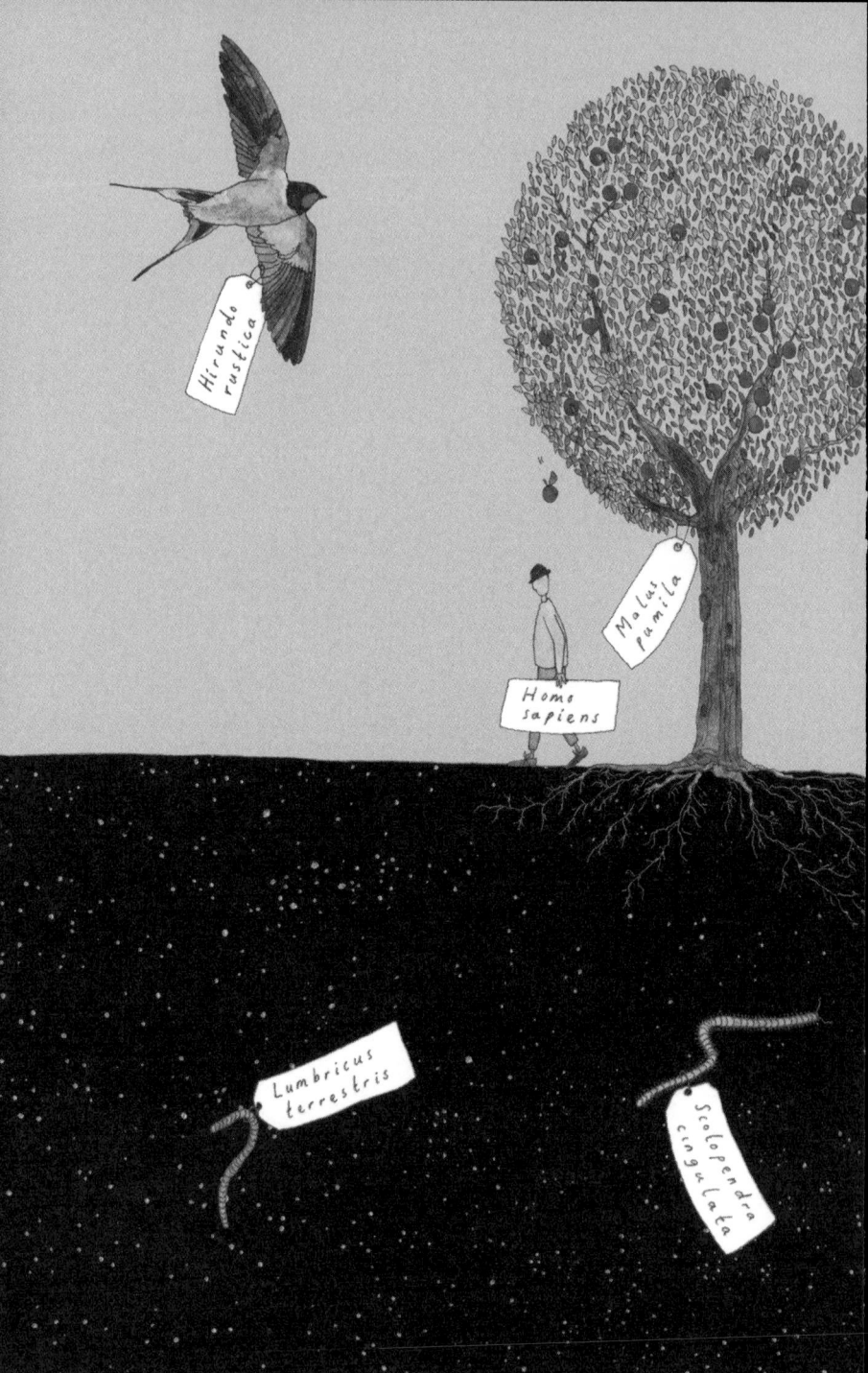

CLASSIFICATION

All living organisms can be placed into groups, based on simple, shared characteristics, and once within those groups it is possible to further separate and divide them into more specific and nuanced ones. This is a process known as taxonomy (coined from the Greek *taxis* for "arrangement," and *nomia* for "method"). Our current system for the classification of every living thing is based on Swedish botanist Carolus Linnaeus's 1735 *Systema Naturae*. Although we have kept the term "Linnaean," which now refers collectively to several separate fields of classification, the system has changed hugely over time—his initial idea for a "mineral" group was abandoned posthaste.

Today, we classify every organism at different levels: domain, kingdom, phylum, class, order, family, genus, and species (species being the smallest and most precise category). We, for example, belong to the class Mammalia, the order Primates, and the family Hominidae. These classifications then lead to the individual and scientific two-part names given to each living thing, known as "binomial nomenclature," which provides the genus and species groups and is always written in italics, with the first word capitalized, the second in lowercase. In our case, it's *Homo sapiens*.

The Linnaean method of classification centers on the similarities between things, rather than on phylogeny or evolutionary relationships, although there are other systems, like cladistics, which are based on genetics and traits that can be traced back to a most recent common ancestor (see page 29). Systems such as cladistics do incorporate more traditional taxonomy, but their primary aim is to reconstruct an evolutionary history. It is perhaps interesting to note that the classification of both new and old species often leads to Latin-stuffed arguments between scientists, because many species will sit just inside or outside certain definitions, shying away from certain criteria just enough to avoid being definitively labeled.

Disagreement aside, it is clear that we have an inherent and almost frantic compulsion to label things—we need to have words for them before we can fully understand, to have a then before we can have a now, to give us any, some, or all control. But naming things also enables better communication, transforms the familiar into the well-worn, and helps us to maintain some semblance of sanity. We are social by nature, drawn to order and to things that save us time.

Having concise terminology for the known world frees up more room for us to fill with dreams.

DAYS AND YEARS

There are years that we remember and years that we seem to forget entirely. With forty or so different calendars in use around the world, and not one of them matching up perfectly to an astronomical year, it's no wonder we are all clinging somewhat haphazardly to the apparent passing of time. (An astronomical year, also called a "tropical," "solar," or "equinoctial" year, is the time taken for Earth to orbit once around the sun.)

The most widely used calendar today is the Gregorian one, introduced in 1582, though in terms of length, it only differs from the Julian calendar that preceded it by eleven minutes. After one thousand years of the Julian calendar, things like equinoxes, solstices, and Easter had gotten horribly out of sync (Easter is what the Gregorian calendar was most preoccupied with fixing), so the length of a year was changed from 365 days and 6 hours, to 365 days, 5 hours, and 49 minutes. While most of us have been using the system ever since, the adopting of the comparatively newfangled Gregorian plan required accommodating the strangeness of leap years, which occur every four years but under Gregorian rules now have to be skipped every one hundred years unless the year is a number divisible by four hundred (a minor detail). In order to complete the change, we also needed to delete ten days from history, and so the days between October 5th and October 14th, 1582, have never actually existed (another minor detail).

The ancient Egyptians are credited with being the first civilization to split a day into twenty-four sections, though as they lacked any fixed measure of time, these hours would vary hugely over the course of the seasons—the length of them would shrink and grow depending on darkness, daylight. The ancient Greeks then proposed equal parts of a day, but even after that, most

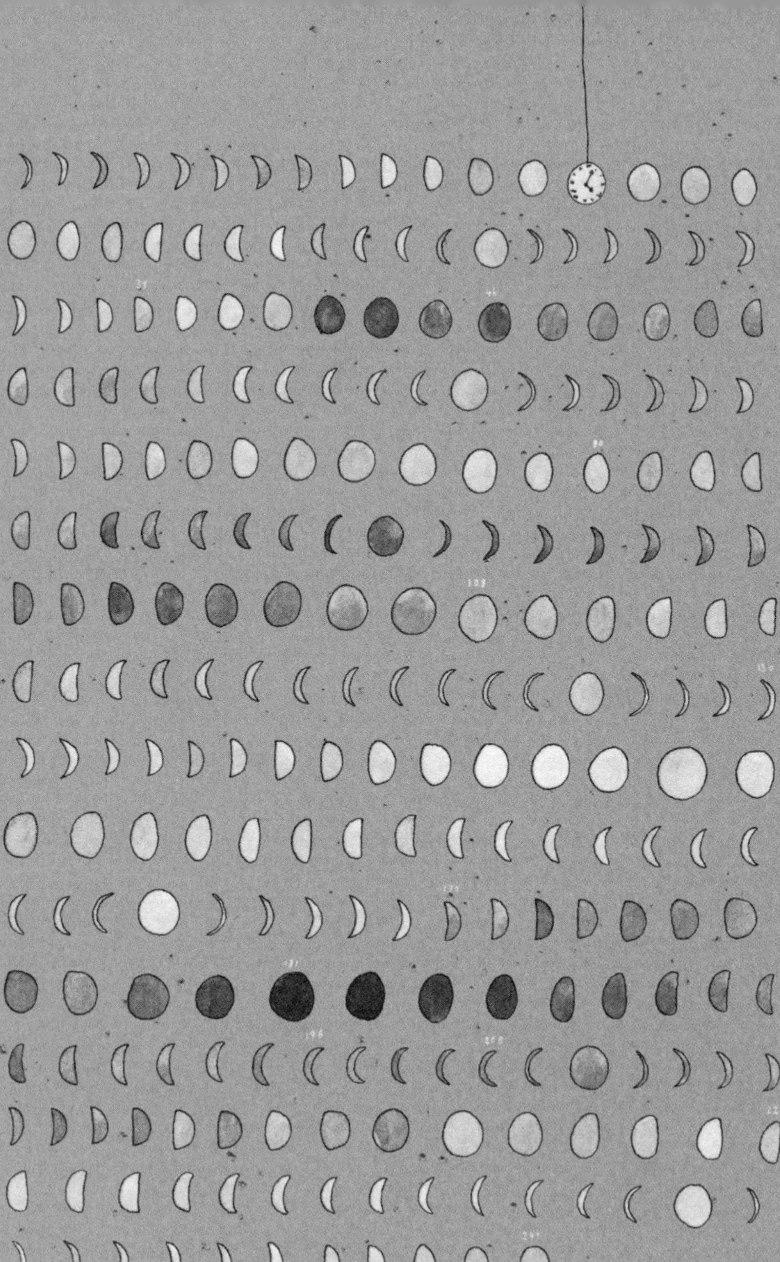

people continued to keep seasonal hours until the introduction of the pendulum clock in late 16th-century Europe.

In 1967, helpfully (or not helpfully, depending on how you want to look at it), the atomic clock became the most accurate keeper of time in human history, and since then the International Bureau of Weights and Measures has defined the second as the duration of 9,192,631,770 cycles of radiation corresponding to the transition between two energy levels of the caesium-133 atom. Such clocks will not lose seconds for billions of years. And yet, we must still go around adding leap seconds to Coordinated Universal Time (UTC) in order to keep atomic time in agreement with astronomical time, and around eight minutes per decade will contain sixty-one seconds, rather than sixty.

If it doesn't make sense here, it might make sense there.

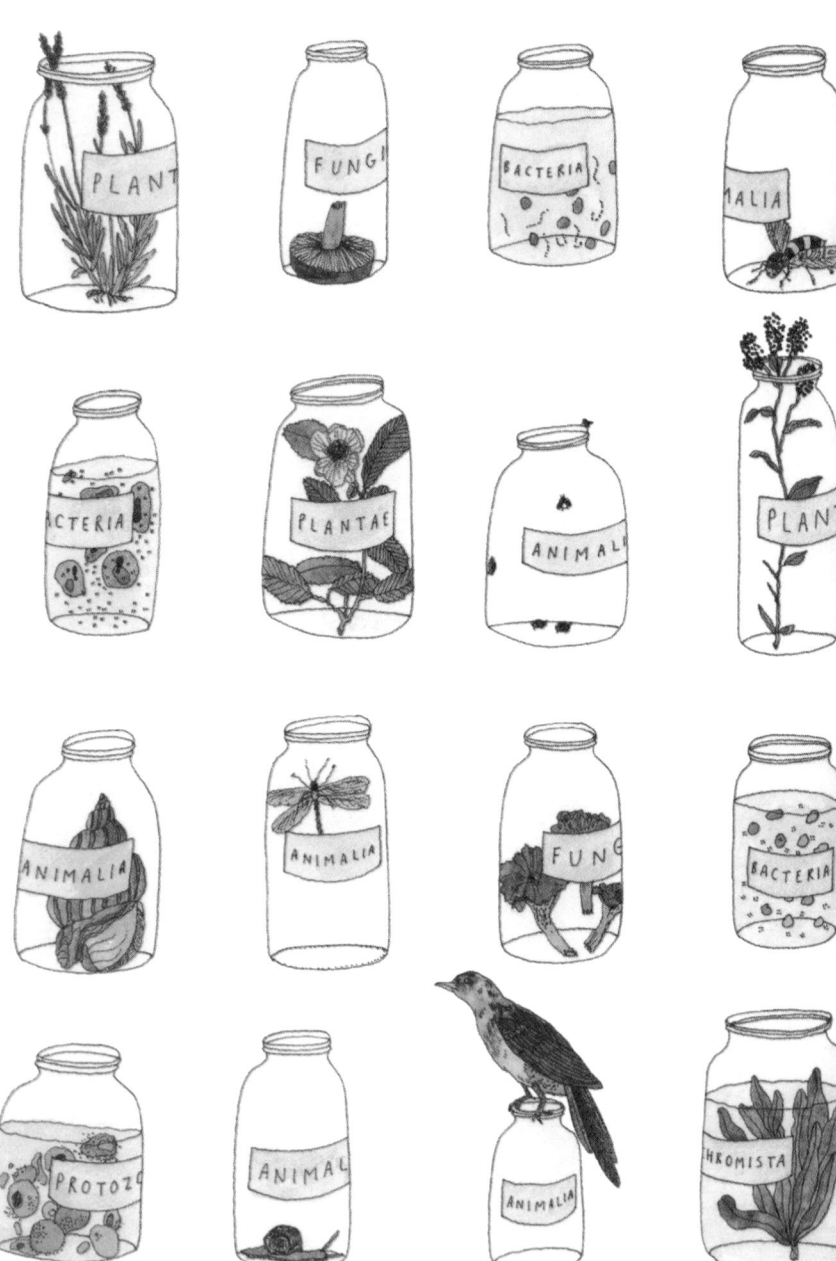

KINGDOMS OF LIFE

After it became apparent that simply putting all living things into just "animal" and "vegetable" kingdoms wouldn't do justice to the endless diversity of life, biologists took it upon themselves to further separate out the system, before settling upon a handful of categories that could encompass everything.

Although there is a category above kingdom, domain, which divides living things into either Bacteria, Archaea, or Eukarya, kingdoms were the most general way you could classify organisms until the 1990s. However, both before and after this time there lies a fair amount of confusion and disagreement regarding similarity and ancestry and all sorts, and we still haven't quite sorted it out yet. (An overview of classification can be found on page 47.)

In 1998, and after much to-ing and fro-ing, an English biologist called Thomas Cavalier-Smith published a new version of the six-kingdom model, which subsequently got changed several more times between then and 2015. However, we have seemingly paused for now on the following seven kingdoms: Bacteria, Archaea, Protozoa, Chromista, Plantae, Fungi, and Animalia.

Humans of course sit inside the last of these kingdoms, Animalia. "Animal" comes from the Latin word *animalis*, which means having breath, having soul, a living thing (I'll be sure that we exist if you are). When applied in a biological context, "animal" refers to all creatures—locusts, a magpie, lizards, humans. But when we use the term in everyday life, we so often and so neatly cut ourselves out of the picture, and "animal" is left to refer to everything except us, indicating only other mammals, or only those with a backbone (even though vertebrates only account for 5 percent of animal species).

We like to point out the differences. We are different from single-celled bacteria because we have many; we don't have rigid cell walls so we can stand across the room from plants, algae, and fungi, who do. We can't produce our own food but instead have to consume it, feed directly or indirectly on other living things. We can move unaided, go almost anywhere we want to, and don't have any need to fear predators. One might suppose such differences must surely make us all terribly special and important. Not in the least.

WHAT EXACTLY AM I
BREATHING IN

While many assume "air" to just mean "oxygen," that only accounts for 21 percent of what you're breathing in. The rest of your lung soup is primarily nitrogen (78 percent), along with some other more unusual ingredients: other gases, contaminants and air pollutants, water molecules, dust, microbes, and spores from plants. You are also inhaling cosmic dust, the leftovers from things broken down in the outer atmosphere, which reach your insides in tiny yet massive amounts: you will definitely be breathing in particles from a meteoroid at some point this year.

All of the molecules in the air are constantly colliding with one another at hundreds of miles per hour, and within weeks can have traveled around the world and been distributed throughout Earth's lower atmosphere. This means that each and every particle has the potential to end up in your neck of the woods, in your lungs.

Within a twenty-four-hour period, the average human will inhale about 9,000 liters of air, taking around 24,000 breaths per day, and more than 8 million a year. Someone who lives until the age of eighty may have taken more than 700 million breaths in their lifetime—unimaginable, wonderful, how easy it all seems.

You rarely ever need to think about breathing, and it would be highly unusual for your respiratory system to ever forget what to do—your lungs, heart, and all the rest of it carries on without you really noticing. As it turns out, it might be worth thinking about. While breathing has long been thought of as an automatic process, one driven by the part of the brain that controls life—inherent things like heartbeat and sleeping

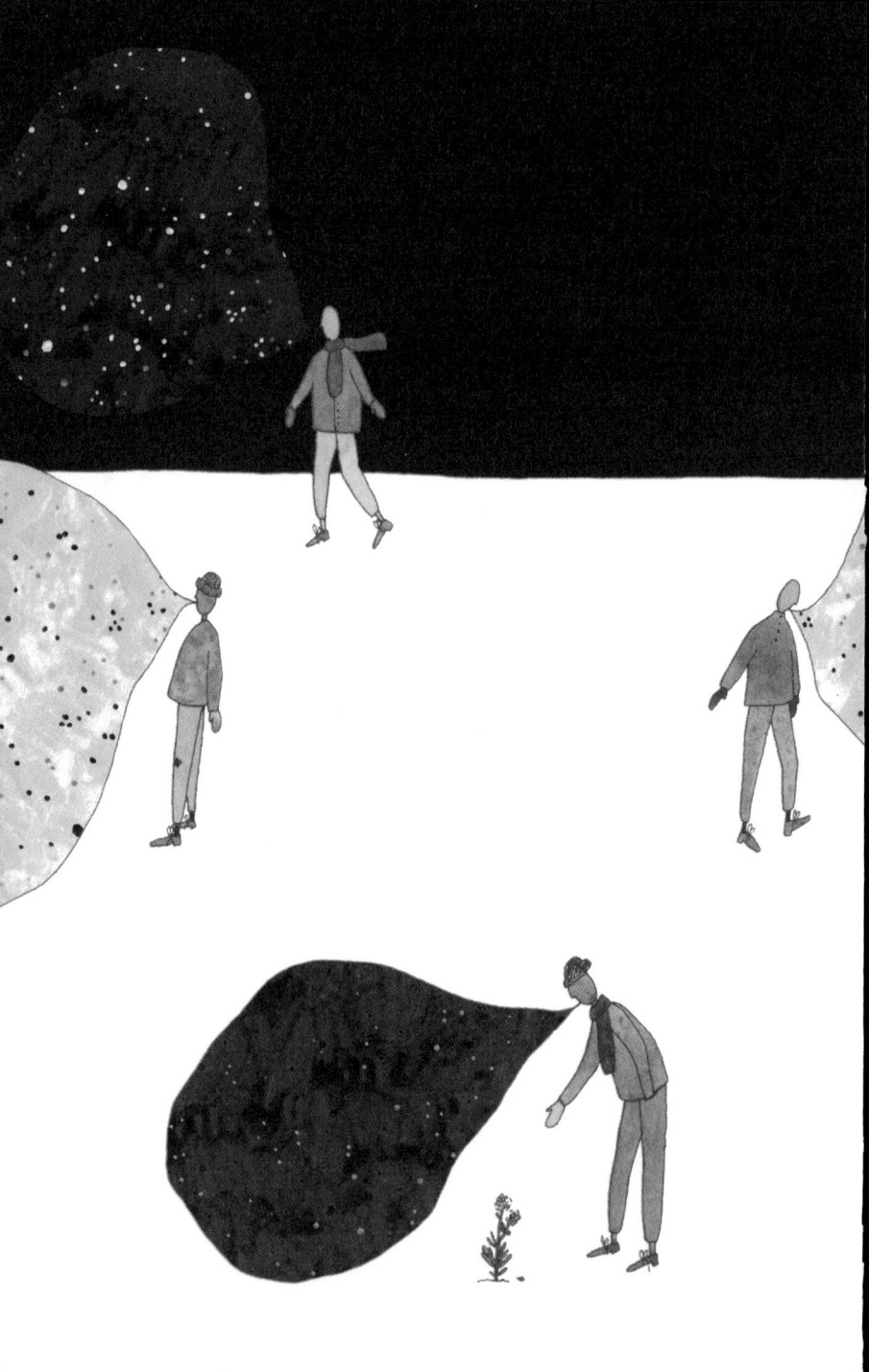

patterns—it can actually change your mind. Breathing at different speeds, even paying less or more attention to the breath, has been shown to engage different areas of the brain. And we are among the only animals who can actually alter and regulate our breathing at will, consciously, rather than it simply happening in response to things such as running, resting, panic.

Everything is fascinating, so don't hold your breath.

YOU'RE THE ONLY ONE I WANT
TO TALK TO

The motion of the Moon, something known as lunar theory, drove astronomers slightly mad for centuries; Isaac Newton said that it was the one problem that gave him severe headaches. There are many irregular aspects to its orbit, its eccentric longing, known as "perturbations," which have kept people busy—and we still don't even know exactly where the Moon came from. While it has a two-thousand-year history of investigation, we can now map lunar motion to a very high level of accuracy.

Ours is the fifth largest moon in the solar system, but there are many, many others—Mars has a couple, Jupiter has at least sixty-nine. But ours is so companionable, so close, and it has a capital letter: it is Moon (if only because nobody knew or noticed for a long time that there even were any others, until Galileo found four moons around Jupiter in 1610).

The influence of the Moon is far-reaching. Because of it, Earth has had an incredibly steady climate (at least compared to a lot of space), and its stabilizing effect on our rotation has been tremendously helpful, not to mention the fact that without it, evolution could well have played out very differently. If Earth had been alone and without lunar tides, for example, there would not have been the same dramatic fluctuations along coastlines, changes which could well have pushed along the evolution of very early biomolecules.

While the Moon may be moving gradually away, the accumulative effect of lunar tides over millions of years means that we currently share a synchronous rotation—a movement similar to that between two dance partners. The Moon rotates at the same speed it revolves around Earth (it's effectively been captured by us), which is why we see the same side of it all the time—a "despun" moon.

It is also there and reassuringly present, always, much like the stars. We can't see the stars during the day because the sky is too comparatively bright, but the Moon can sometimes appear even brighter during the day than at night. How glad we can be, that we have someone to figure out this universe business alongside, to dance with, to gradually lengthen our days and keep us slow.

LET SLEEPING MOUNTAINS LIE

Most mountains on Earth are the result of tectonic plates careening impolitely into each other, disagreements left unresolved for millions of years. Tectonic plates, the sublayers of Earth's crust, are responsible for the drift of continents, earthquakes and volcanoes, deep ocean trenches, and entire mountain ranges—folded up dramatically and seemingly as easily as paper.

There are only fourteen mountains on Earth with peaks standing higher than 8,000 meters (about 26,300 feet), the tallest of which is Mount Everest, lying in the Himalayan mountain range, which sprawls across Asia from Afghanistan through Pakistan, India into Nepal, Tibet through Bhutan, before finally tailing off in Myanmar. Although there are volcanic mountains, Everest is not one of them, and the range was formed by the collision of two continents 40 million years ago—the top of Everest is softer, sedimentary rock, formed from the skeletons of creatures who were minding their own business in a warm ocean. Its Nepalese name is Sagarmatha, which means "mother of the universe," and it still grows a little each year.

While there is no universally accepted definition of a mountain, most geologists classify them as such if they rise at least 300 meters (1,000 feet) above the rest of the surrounding area. They cover 20 percent of Earth's surface, and are home to about 10 percent of the world's population—we sleep on our sides on theirs. Most of our rivers have their unassuming sources in mountains, in thin air, and over half of humanity depends on them for water.

As hard as it may be to believe, mountains used to be considered ugly and foreboding by the vast majority of people. It wasn't actually until the end of the 18th century, when the Romantic writers began to get a bit carried away, that we started to marvel

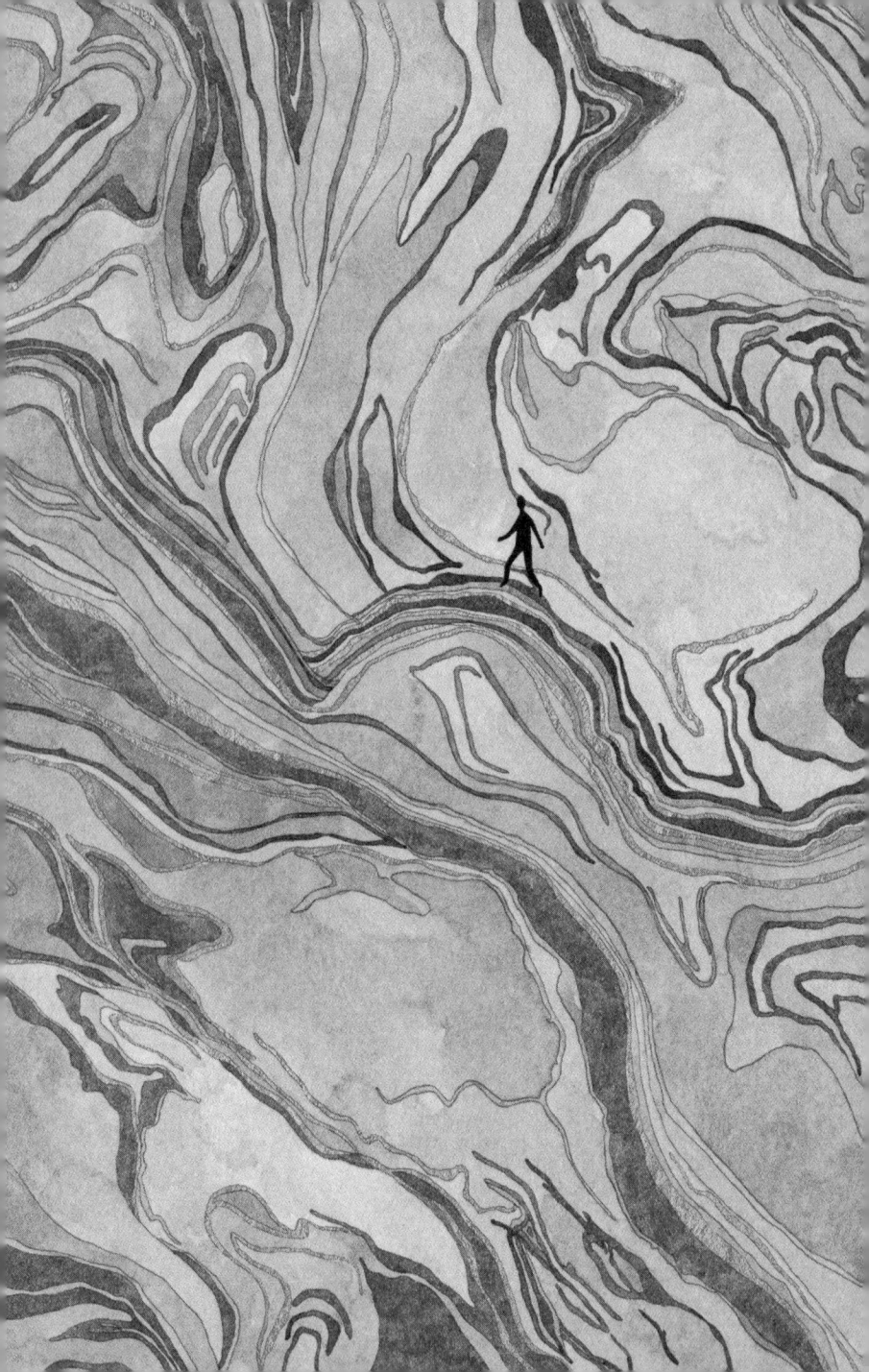

at their existence, collectively excited by the thought of summits. Today, we can also marvel at how far-reaching and dreadful the consequences of our chaotic existence on the planet are. The melting of glaciers in mountains, caused by global warming and the rise in the average temperature, is having an effect on the very spin of Earth—affecting the movement of our planet on its axis, as the weight of the water previously contained within these ice sheets is redistributed from higher to lower latitudes.

It is precisely this sort of reality that one cannot ignore—hold on, let go, hold on again. Be one of the ones who doesn't stumble about with eyes closed and hands in pockets.

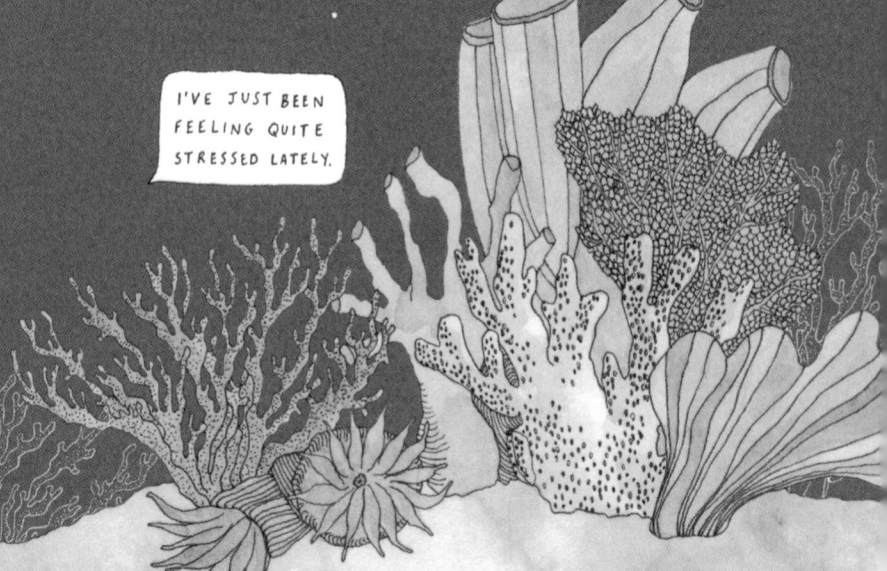

STRESSED - OUT CORAL

Although it can be hard to comprehend when looking at them, corals are actually not plants, but animals. They do however rely almost entirely on photosynthesis for their existence (see page 3), and the microscopic algae living in their cell linings respire in this way, providing a coral with 90 percent of its energy. It is for this reason that reefs develop best in shallow waters, where the water is warmer, brighter, more of it bathed in sun.

Coral reefs cover less than 1 percent of the ocean floor, yet are one of the most diverse ecosystems in the world and have a huge impact on both the surrounding water and the surrounding land. Among other things, they protect coastlines from tropical storms and the associated erosion, facilitate the existence of thousands of other species, and are even important when it comes to our medical research and the development of treatments for diseases such as cancer and Alzheimer's.

Reefs need to maintain a delicate balance, and our loud activities on the planet have, one way and another, led to large amounts of damage. Corals grow incredibly slowly, between 2 millimeters and 10 centimeters each year, and they have been unable to recover fast enough to keep up with the rise in ocean temperatures and acidity, destructive fishing practices, pollution, and careless tourism. Roughly half of our coral reefs have been lost in the last thirty years. When conditions in the water become too much for the corals, they force out the life-giving algae in their cells as a natural response to overheating, and then turn a sad shade of white. This coral bleaching doesn't necessarily kill them, but it can take many, many years for them to recover and even if they do, their reproductive capacity is likely forever affected.

But there is much that we can do to help slow the rate of destruction, and steps we can take to help encourage the rehabilitation of coral reefs. In 2017, scientists played matchmaker in an area of the Great Barrier Reef, harvesting coral larvae and growing millions of them in tanks, before returning them to the ocean floor in protective underwater nets, with many continuing to grow successfully. A similar project took place in the Philippines, where fishing techniques that use explosives had damaged huge areas of coral reef, and endeavors such as these are immensely important, because we simply cannot afford to lose such an ecosystem entirely.

It is something worth worrying about, something worth acting upon. Along with all of the other things you need to worry about and act upon, of course.

DANCING IN EMPTY SPACES

Humans have a tendency to think that we're all that, when in fact, we're not much of anything at all.

While the mass of an atom comes from its nucleus, that doesn't exactly take up much room, and the 7 billion billion billion atoms making up our bodies, along with all the other atoms in the universe, are actually 99.9999999 percent space. This space isn't exactly empty, at least not in the way that you might suppose, but instead is filled with electrons refusing to get overly close to one another, as well as wave functions and invisible quantum fields and ideas too huge to fit on a single side of paper. If you were to take away this "space," you would be able to fit the rest of you into a cube less than 1/500th of a centimeter in width.

The nucleus within an atom is about 100,000 times smaller than the whole structure (see more on page 17), so much smaller that it would be like a single fly inside of a cathedral. Surrounding the nucleus is a cloud of electrons, which are often depicted in science textbooks as small orbs circling the nucleus in a very orderly and matter-of-fact way. In reality, these electrons are more like a giant swarm of birds—it isn't possible to observe the exact movements of an individual but you can see the entire flock in motion.

There is really no better word to describe what electrons do than dancing, and it's not embarrassing or random dancing either; they follow a beautiful series of patterns and steps that were laid out by a single mathematical equation, one named after the Austrian physicist Erwin Schrödinger, who did extraordinary work in the field of quantum theory. These dance steps vary, and the electrons never tire, and no two will follow exactly the same steps, something known as the "exclusion principle."

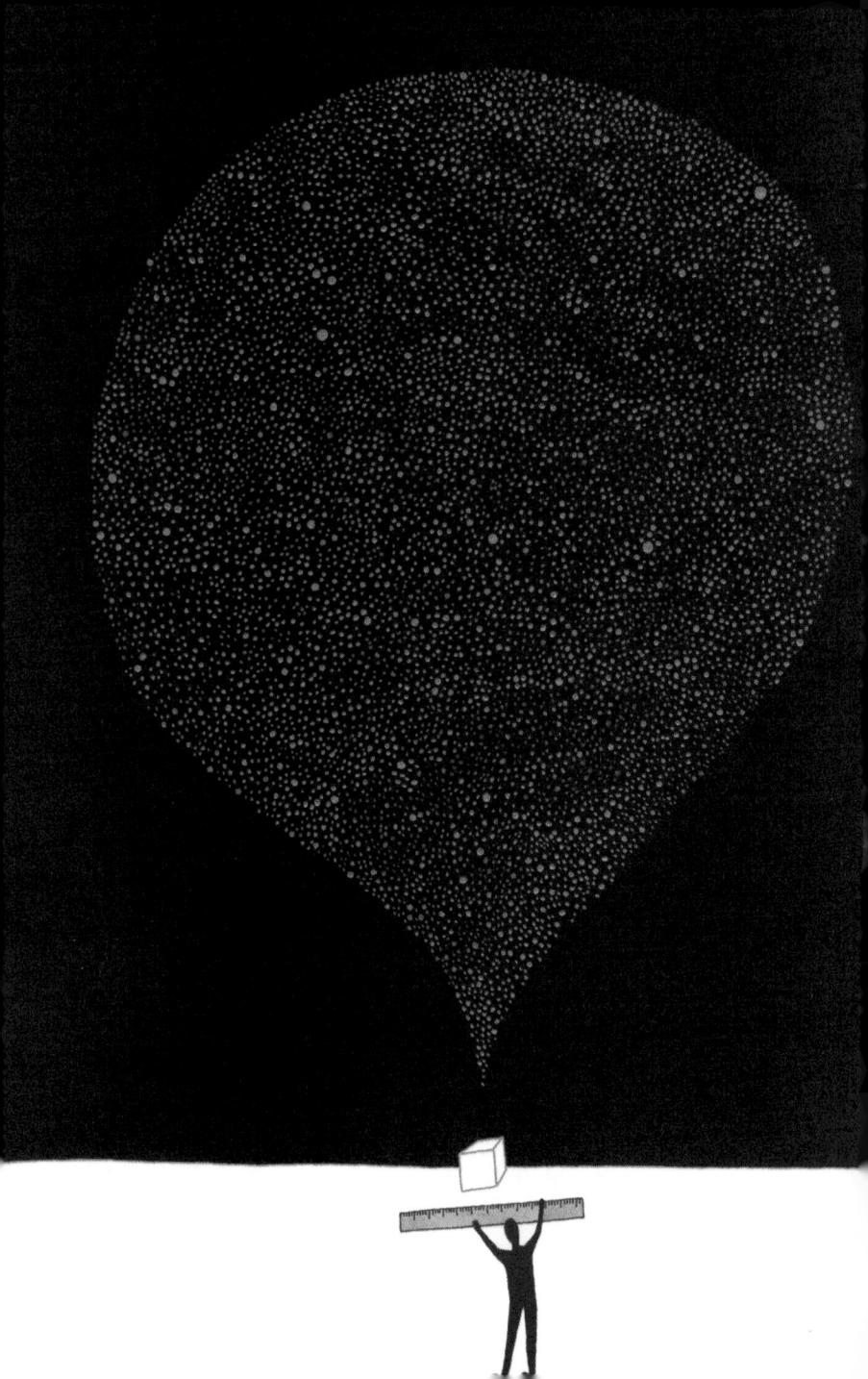

You appear, at a subatomic level, to be dancing all the time, which might provide food for thought next time you feel unable to do more than just a little self-conscious swaying, or an eyes-on-the-floor foot shuffle. Although, saying that, your body replaces 98 percent of your atoms every year, so maybe don't get too attached.

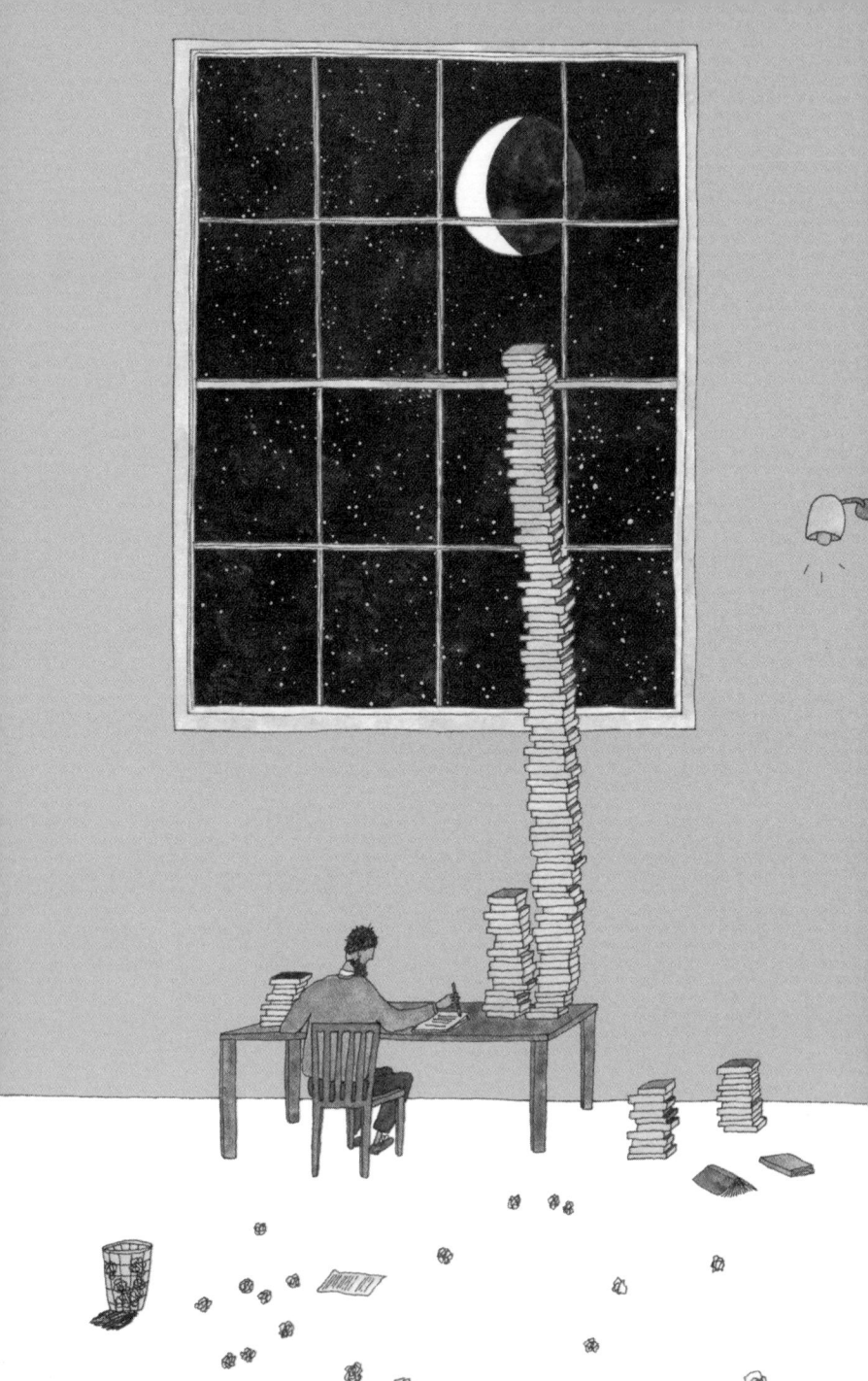

THEORIES ARE NOT GUESSES

There is a fair amount of misunderstanding when it comes to the terms scientists use to explain and organize things. The words they employ are often the same as those we use somewhat thoughtlessly every day, but the definitions can differ wildly. For instance, for most nonscientists, the word "theory" would likely mean a guess, a hunch, some vague speculation about something or other. Not so in science, where "theory" refers to an incredibly well-established explanation, a scientific theory, one that is unlikely to be altered much by new evidence, that has been confirmed repeatedly through observation, or experiment.

Good examples of scientific theories include the big bang theory, Darwin's theory of evolution, Einstein's theories of relativity, and the theory of everything. These theories cannot normally be reduced to one sentence or statement, or one neat and tidy equation, but they do represent, often with astonishing clarity, fundamental things about the ways in which nature works.

Another common misconception is that theories turn into laws when they have been subjected to enough research, enough evidence, and time. This isn't at all the case, and scientific laws and scientific theories are not in any way interchangeable. A law in science is the description of an observable phenomenon, but it doesn't explain why it exists. It's a scientific theory that attempts to cleanly provide the most logical explanation as to *why* a phenomenon is the way that it is. To put it simply: laws predict what happens in the universe; theories propose why.

It is perhaps also worth noting that hypotheses are not guesses of either a wild or an educated nature, not stabs in the dark. When scientists form hypotheses, they are basing them on existing scientific knowledge, on prior experience, on observations and logic. They are the proposed explanations for a fairly narrow set of happenings, while theories are broader, consisting of one or more hypotheses, all of which would have been exposed to rigorous and repeated testing.

Although a scientific theory being overturned completely is a near impossibility, even small disproved pieces of one can be hugely valuable, as these holes then frequently lead to newfangled discoveries, ones that were previously unimaginable. Because each reworking of a theory will contain more knowledge, more analytical oomph than the last, our understanding of things can become more accurate and embellished over time.

That is really what science is—fluctuation, a process of reconfiguration of observable truths, something that challenges our most deeply held assumptions and wreaks a delightful havoc with common sense.

THE UNIVERSE IS OLDER THAN YOU

The age of the universe as we know it is 13.8 billion years, give or take 120 million, so it has had an awful lot of anniversaries since the big bang, which is the only thing we can reasonably refer to as a beginning. Light has been traveling through it for all that time, so the most distant and outer parts of the universe we can gawk at are only ever as much as 13.8 billion light-years away—what we call the "observable universe" is really just a small splinter, and it likely extends in all directions, indefinitely.

It's possible to know roughly yet precisely how old the universe is by doing things such as looking at the oldest objects in it, or measuring the rate of its expansion. Logically, the universe cannot be younger than anything in it, and the globular star clusters that orbit the Milky Way, the oldest things we know of, are between 11 and 18 billion years old (so a minimum age of 11 billion years can be imposed on the universe). And measuring the expansion of the universe today allows us to work out how it has expanded in the past (going, in theory, infinitely backwards). Though this figure of 13.8 billion has been reached using measurements from various sources, the changing density of matter in the universe over time is one of the most helpful ones, as a universe with a low density of matter is older than one with a high density—today, the universe contains only 4.9 percent normal matter (the rest being 68.3 percent dark energy, 26.8 percent dark matter, and only a tiny amount of anything else, such as neutrinos, photons, and radiation).

Extraordinary, that the universe can be dated with 99.1 percent accuracy, when even ten years ago the words "precise" and "cosmology" would likely not have even been sitting in the same room, let alone having a conversation.

IT DOESN'T
MATTER.

OH VERY
FUNNY.

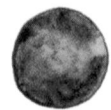

You, you are old too. Many of the atoms in your body, like the carbon and the oxygen (see more on page 1), are brewed inside giant stars, and so you are a slightly different configuration of all that there ever was going to be. Having been around in some shape or form for those 13.8 billion years, it's no wonder you're occasionally exhausted.

And you had to be you, you are the only you who could possibly have been—here, then, now. Everything that has ever occurred since the big bang has happened according to the laws of the universe, and they are just that, *laws*. So while we may not know about all of them, or even understand completely the ones we do know about, it is almost as if it was all already decided upon: all those happenings and events unfolding in the only way that they possibly could have.

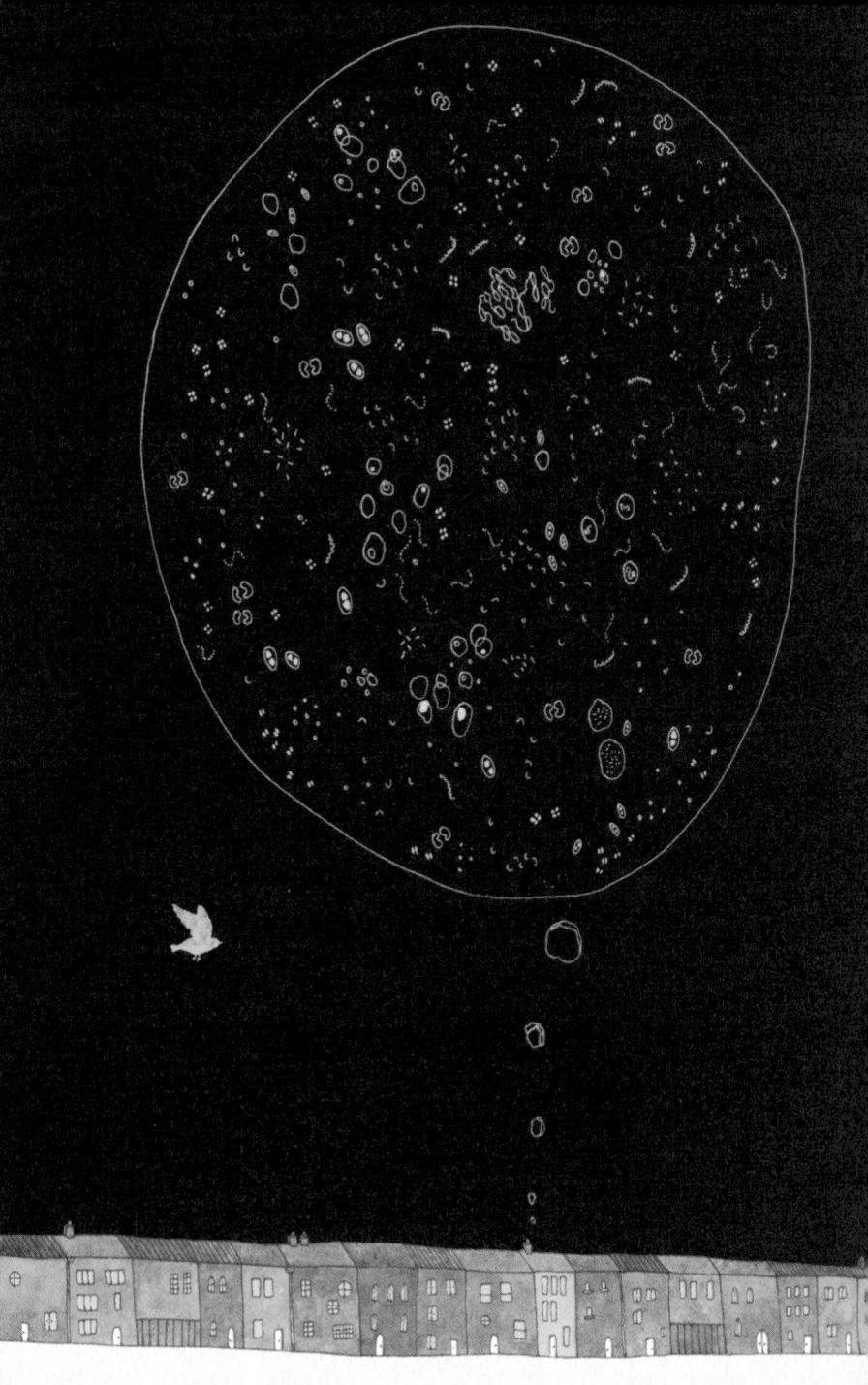

YOU ARE MOSTLY BACTERIA

While it has been reprinted and repeated with wild abandon, the fact that bacteria outnumber the rest of the cells in your body by about ten to one, the ratio is likely a much less dramatic one to one—that is, one part you, one part bacteria. There are, however, plenty of factors that can affect this ratio, things like body weight and how feisty your immune system is feeling. It's estimated that the cells in your body number somewhere between 15 and 724 trillion, while the single-celled microbes, the bacteria, number between 30 and 400 trillion. Perhaps more remarkable than that is the fact that scientists still remain unsure about the exact amounts of these things. Your body is a microbial wonderland.

Bacteria are one of the earliest life-forms, and while it can be both unnerving and peculiar to know that you're covered in them from head to toe (the singularly alarming part can be when you see images of them under microscopes), the vast majority of them do you a tremendous amount of good: keeping harmful bacteria at bay, controlling the surface environment on your skin, and playing a crucial role in the development and maintenance of your immune system. If all that sounds lovely, it's because it is.

Perhaps the greatest achievement of bacteria is the gut, part of an organ system that contains the largest number of bacterial species found anywhere in the human body—an incredibly complex community of microorganisms, known as the gastrointestinal microbiota, or gut flora. The relationship between a person and their gut flora is not only mutualistic, but remarkable, and it is only in recent years that it has begun to dawn on scientists just how complex and important the relationship between the gut and the brain is.

The two are connected by an extensive network of neurons (nerve cells), chemicals, and hormones, collectively called the "enteric nervous system," which stays in touch with the central nervous system (that which links the brain and the spinal cord) but can also act independently of it. It seems that the bacteria in your gut are capable of wielding power over more than just your breakfast—they can influence your perception of the world, your behavior. Most of the neurons involved in brain-gut discussion are carrying information to the brain, not receiving it, and it's strange and wonderful to think that your gut might have more of an effect on both your long- and short-term mood than anything else.

Language seems to have suspected this for longer than science—gut feelings, go with your gut, gut-wrenching—and it would surely be interesting to see what happened if we listened to our bacteria rather than our brains for a while.

YOU ARE ONLY REMEMBERING THE
LAST TIME YOU REMEMBERED

For a very long time, it was believed that long-term human memory was comparable to a library where things, events, and occasions got placed on shelves, some being reread very infrequently and others picked up and leafed through often, with most gathering dust but still there if you ever wanted to look something up. In recent years this notion has been entirely overthrown, and we now know that all of our memories are susceptible to drastic alterations.

Each and every time you remember an event from your past, brain networks change it in ways that in turn affect later recollections of that event, so you are never remembering accurately, not even remembering much at all. Memories, as it turns out, are not in the least bit static or unchanging, and your brain constantly rewrites them using current information, integrating factors like environment, time, and mood—such imperceptible revisions all adding, however slightly, to our gradual misremembering of everything. An author of one of the first studies showing these faulty tendencies put it all too succinctly: "Your memory of an event can grow less precise even to the point of being totally false with each retrieval." Be it sad or fortunate, our recollections usually end up more fictional than the stories we tell, more fictional than the books that handed us the memories in the first place.

But there are reasons for these imperfect copies, these slight but repeated shifts within sequences. Your memory is there to help you make good and useful decisions in the present, in current situations, so it needs to stay up-to-date and not linger or

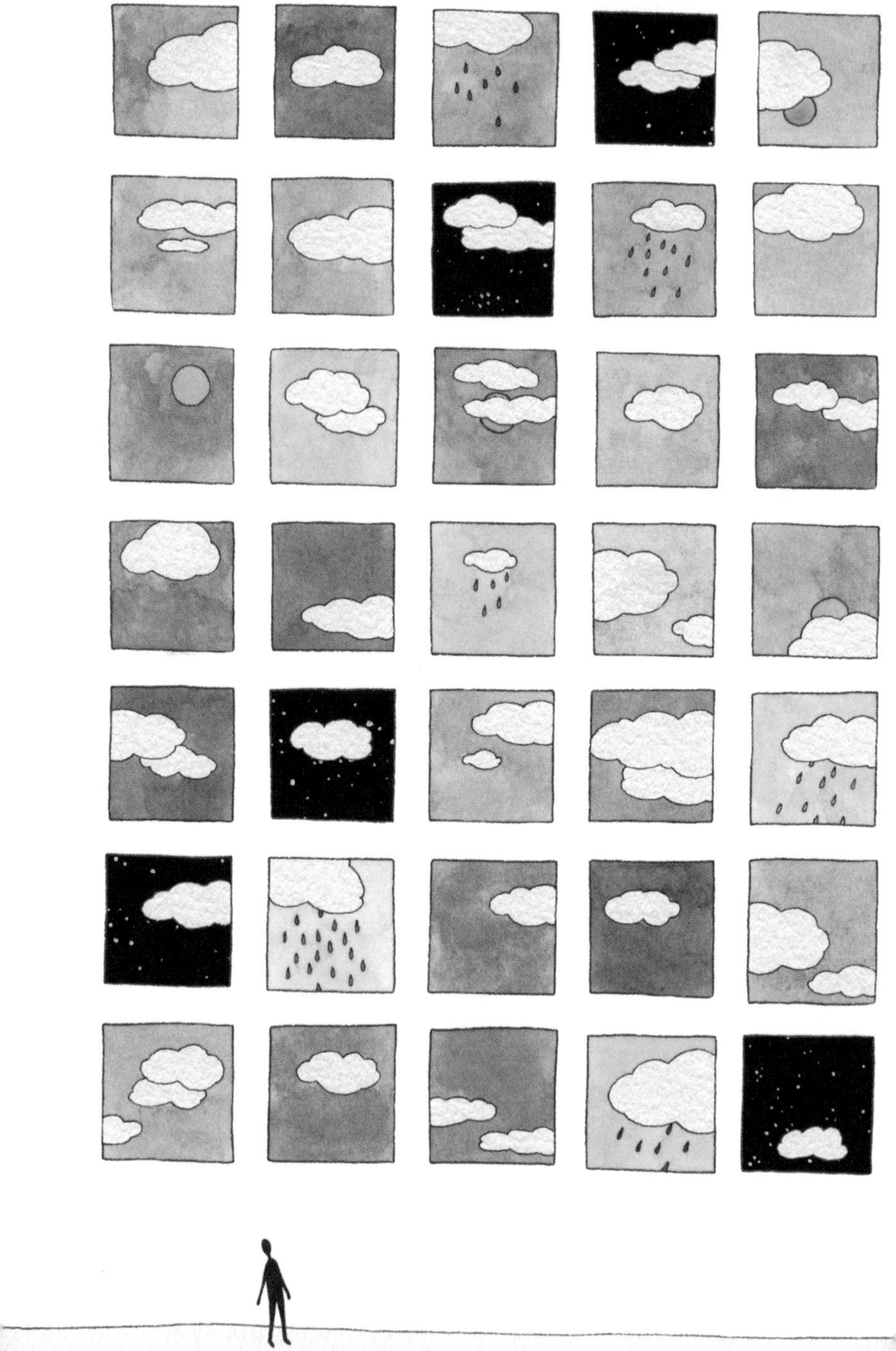

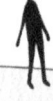

dwell endlessly in the past. Unfortunately, this means we also make for horribly unreliable witnesses, because the brain continually filters memories through who you are now, what you presently think and feel. The memories we hold on tightest to, the ones we revisit the most frequently, are woefully going to be the most inaccurate, precisely because we have gone back over them so many times. And it seems that in order to remember we must also forget; we can retain certain things so long as we suppress others—even then, you can't get caught up on all the details, only a selection of distorted ones.

Our memories are clear, vivid, and, very often, entirely wrong.

THE LANGUAGE OF SCIENCE

Scientific language isn't designed to appeal to human ears, isn't especially melodic; it lacks emotion and freedom of expression, avoids first-person pronouns, and adheres strictly to formal, well-tested rules—heaven forbid you should use an exclamation mark. It is explicit in meaning, and what you see is very much what you get.

Yet scientific language also remains stubbornly inaccessible for most nonscientists, for the reasons above but also because it takes familiar words and puts them in entirely different contexts, as well as introducing a whole other vocabulary that a person would never normally have reason to encounter.

In more ways than one, language remains one of science's largest unsolved dilemmas, as the discussion and publication of research is now nearly entirely in English. Native speakers tend to assume, wrongly, that all of the important information will be in English anyway, while nonnative speakers hesitate to publish their findings in their own mother tongues as it might be overlooked, or the research unnecessarily duplicated. Because of this, we're missing out on new, important studies published in other languages, especially in fields like biodiversity and ecology, where much of the work being done is in far-flung countries.

This is a problem with a history. From the 15th to the 17th century, scientists normally used their native language to discuss ideas, and then published their work in Latin, which sensibly seemed like a neutral ground. But Latin gradually lost its grip in the realm of science; Galileo's first findings concerning Jupiter and its moons were published in Latin, but his later work was in Italian; Newton's later work appeared first in English, whereas earlier ideas of his had initially been available in Latin. By the early 19th century, just three languages accounted for most of the communication and written research: German, English, and French. Now, after centuries of competing for attention, English has definitely become the lingua franca, the tongue of science.

More than anything, this one-tongue-fits-all approach means that other languages are at risk of losing their unique ways of communicating ideas, as they struggle to keep up with the ever-growing and ever-changing vocabulary of science. When such narrow boundaries are placed on thought and discovery and evolution, nothing in particular is gained, and everyone ends up sounding a lot like everyone else.

IT GETS COLDER AFTER SUNRISE

They say that it is darkest just before the dawn, coldest too, but it's technically darkest halfway between dusk and dawn, and as the sun starts to appear on the horizon, it can often be colder than during the thin-feeling darkness that preluded it. On a clear night (free from heavy cloud or any cloud at all), it can be several degrees cooler even an hour after sunrise. You see, morning comes with a delay.

Everything can both gain and lose heat, and logically when more heat is lost than is gained, that thing will cool: your body, celestial bodies, our Earth. During the day, the sun is doing its best to heat the planet up; although Earth also radiates huge quantities of heat outwards all the time, when the sun is up this relationship is balanced, or else more heat is gained than is lost and we all get a bit warm. But after dark, when the sun has slipped from view, Earth continues to put out heat and to cool down; its surface temperature drops, as does the temperature of the air around it, the air that makes you move closer to the person you're walking beside as the streetlights come on.

Then, dawn. We are not surprised that it is here, again, the hazy pinkish light that leaves you feeling unapologetically philosophical and impossibly small. The first rays of sunlight stroking over landscapes and people are weak, they lack the strength to contend with all the heat still being lost from Earth, and so the air temperature continues to fall. Depending on where you are, cloud cover, humidity, and a plethora of other factors, this continuing-to-cool period can vary. It might last minutes in a tropical rain forest, while at the poles this cooling could itself last for several days.

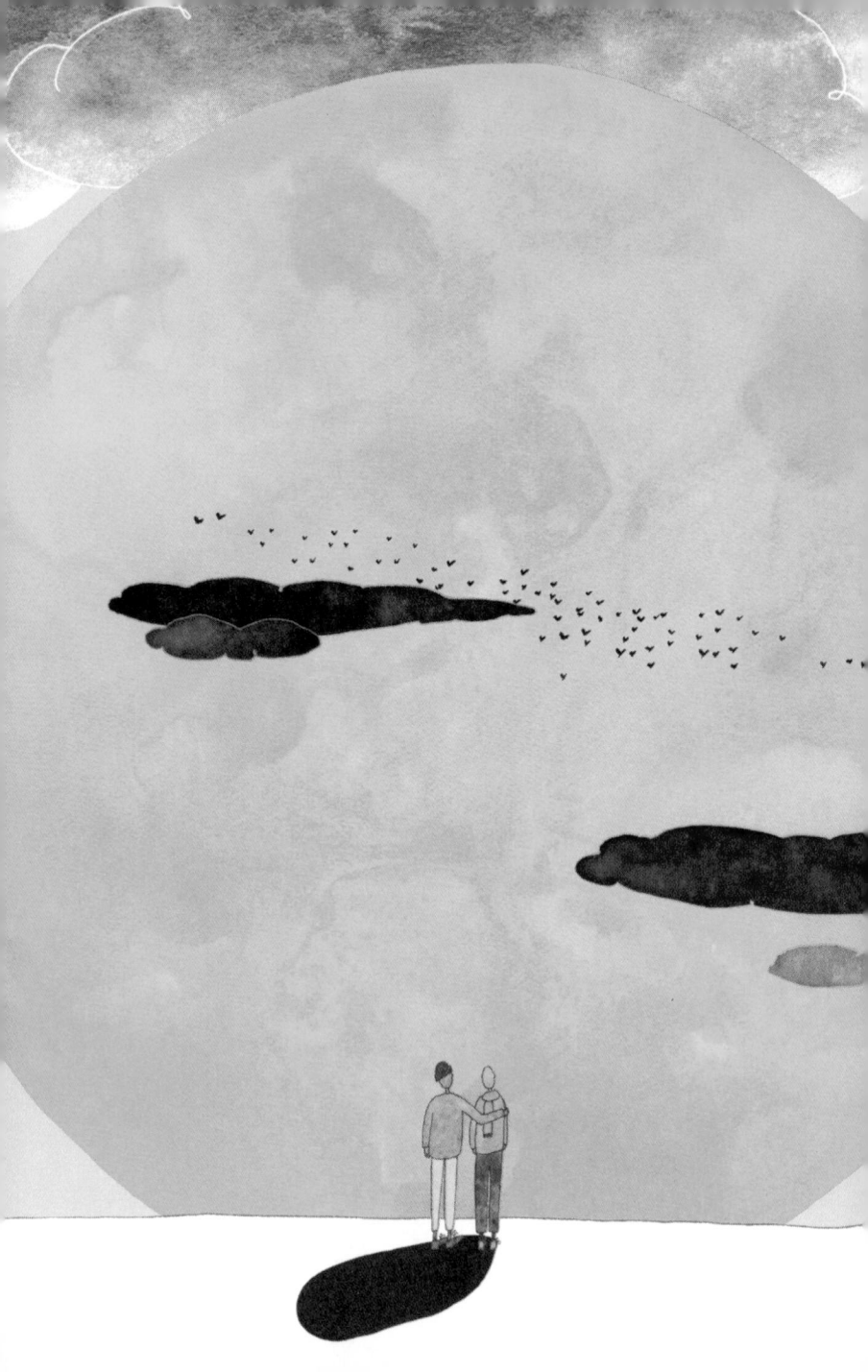

Wait a little longer, though, and the sun will slowly and then all at once become blinding. A thermal equilibrium will be reached, then the ground will begin to warm up, the plants around you will stretch out their fingers, and your hands will no longer be cold.

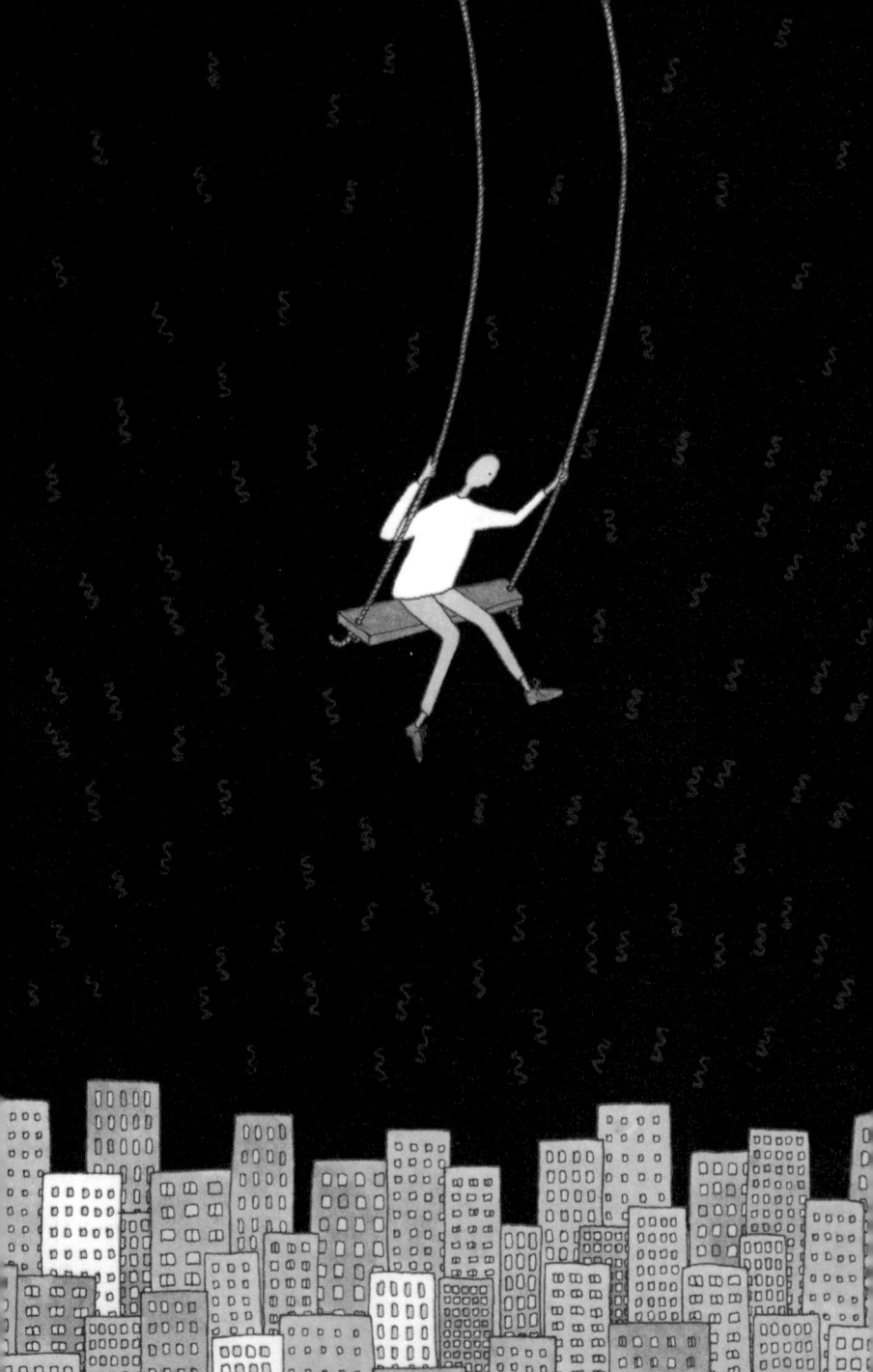

YOU EMIT RADIATION ANYWAY

All things, including you and me, emit radiation, and while for many people the sound of this might be alarming, it is simply a question of how much, and what type. Radiation is really only energy, in the form of waves or particles, passing through either space or materials.

Some confusion and concern likely stem from the fact that the word "radiation" is used for both electromagnetic and nuclear radiation, both of which can be naturally occurring or man-made. Most types of electromagnetic energy are, for the most part, harmless.

You are hit by 15,000 particles of radiation per second, which amounts to 500 billion each year, but nearly all of this is coming from natural sources. And it is likely that you are consuming trace amounts of radiation all of the time; potassium-40, a naturally occurring isotope of potassium, is the largest source of radioactivity in animals, including humans. Bananas and Brazil nuts are two common examples of things that contain higher levels of potassium-40, but you would have to eat six hundred bananas to consume a quantity of radiation similar to the exposure from one X-ray (the average American receives the equivalent of thirty-six X-rays each year from natural sources alone).

It also isn't worth worrying about the radiation from radio waves, which are everywhere; a single candle emits more radiation than your cell phone, so you can go right ahead with your lengthy long-distance telephone calls.

Most of the radiation emitted by human beings is infrared, a type of thermal radiation, part of the electromagnetic spectrum and entirely harmless—nothing more than heat. Warmer objects, logically, emit more infrared radiation than cooler ones, depending on surface temperature, area, and the characteristics of said object. It's most reassuring to remember that while everything on Earth is constantly washed in a radioactive bath of sorts, mostly all of it goes completely over our heads.

IT WAS ONLY A DREAM

There are dozens of different ideas when it comes to the question of why we sleep, but no general consensus, which seems fascinating given that it's perhaps the most important behavioral experience we share—we spend, on average, a third of our lives with our eyes closed.

Two things that everyone agrees on are the stages of sleep we cycle through during a night, and what happens to the body and mind, more or less. The first stage, light sleep, is drifting in and out, commonly accompanied by sudden and involuntary muscle contractions or movements, when you can still be woken easily. During the next stage, eye movements cease and the brain waves begin to slow down, though there are still small bursts of activity in the brain known as "sleep spindles." The third stage brings slower brain waves called "delta waves," and by stage four there is no eye movement or muscle activity—stages three and four together are known as deep sleep, and it's usually mighty difficult to wake someone at this time.

We then enter the phase of sleep known as REM sleep, which stands for "rapid eye movement." Along with this eye movement comes rapid, shallow, and irregular breathing, but the muscles in the limbs are temporarily paralyzed. It sounds unpleasant, but it is a very important and restorative part of the cycle, and the difference in brain activity between REM sleep and non-REM sleep is as marked as the difference between sleep and wakefulness— the vivid and elaborate dreams recalled upon waking have most likely happened during REM sleep. One complete cycle through these stages takes about 90 to 110 minutes, though the amount of time spent in REM sleep normally increases with each cycle.

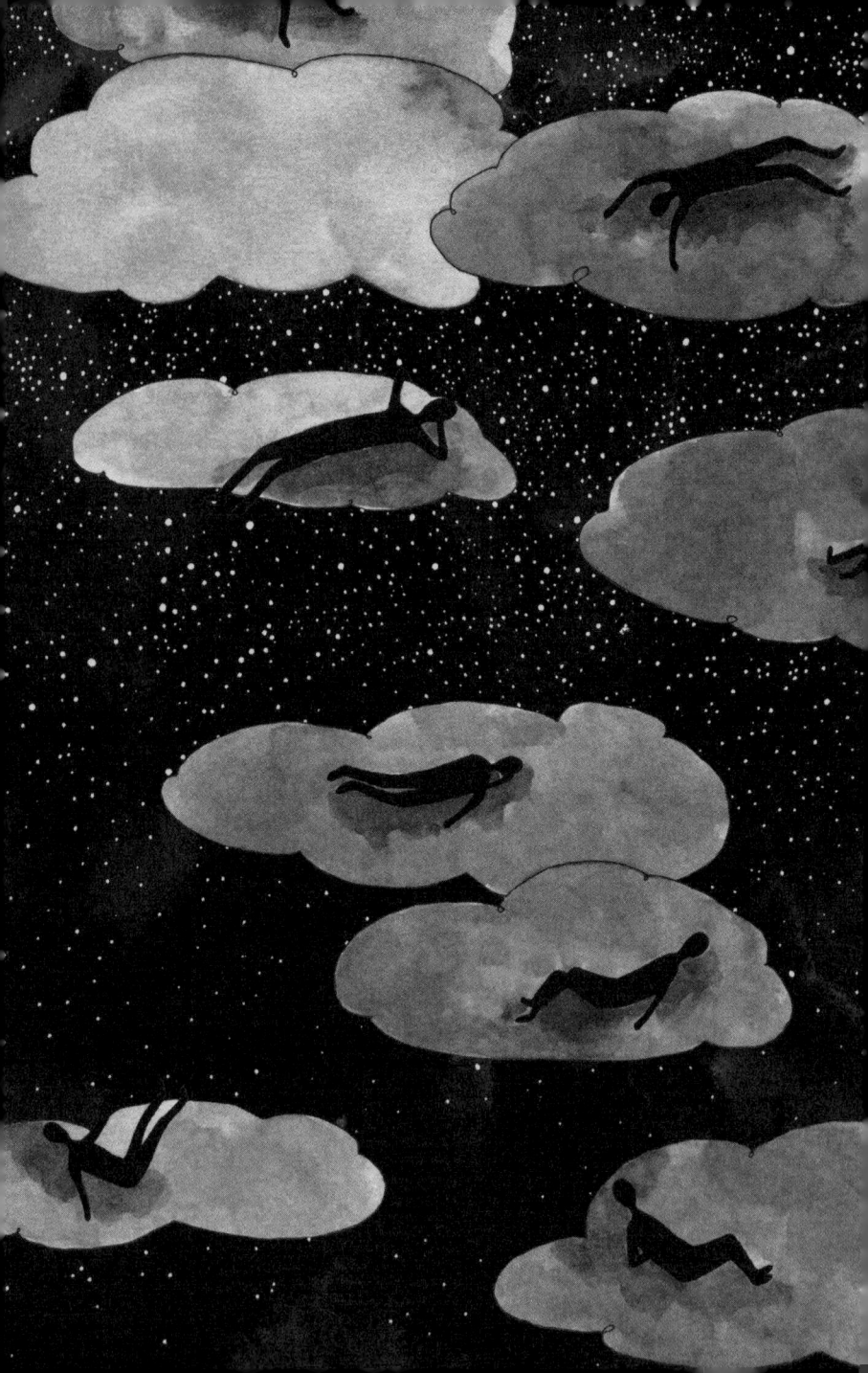

Across the animal kingdom there is huge variation in the duration of time spent slumbering: koalas sleep for twenty-two hours a day, elephants usually get a couple of hours of shut-eye, and we seem to be somewhere in the middle with eight. For us, sleep isn't simply a question of replenishing energy; the energy conserved by sleeping when compared to staying up all night is the equivalent of eating or not eating a small piece of cheese. Some theories about sleep highlight the reversal of damage, the possibility of it being a kind of "cooling down" for the body and mind, and others, like the information processing theory, emphasize the apparent processing and consolidation of memories, the sorting and filtering of experiences from the previous day.

While it remains for now something of a biological mystery, it's very clear how drastically we can be affected by both the lack of it and an excess of it—neither oversleeping nor sleep deprivation will get you very far. In Western societies, chronic sleep deprivation has almost become expected, normal, even admirable, but it has serious long-term consequences for health, both mental and physical. Mysteriously necessary, there seems to be no end to the importance we can place on good rest—as the writer Jim Butcher said, "Sleep is God. Go worship."

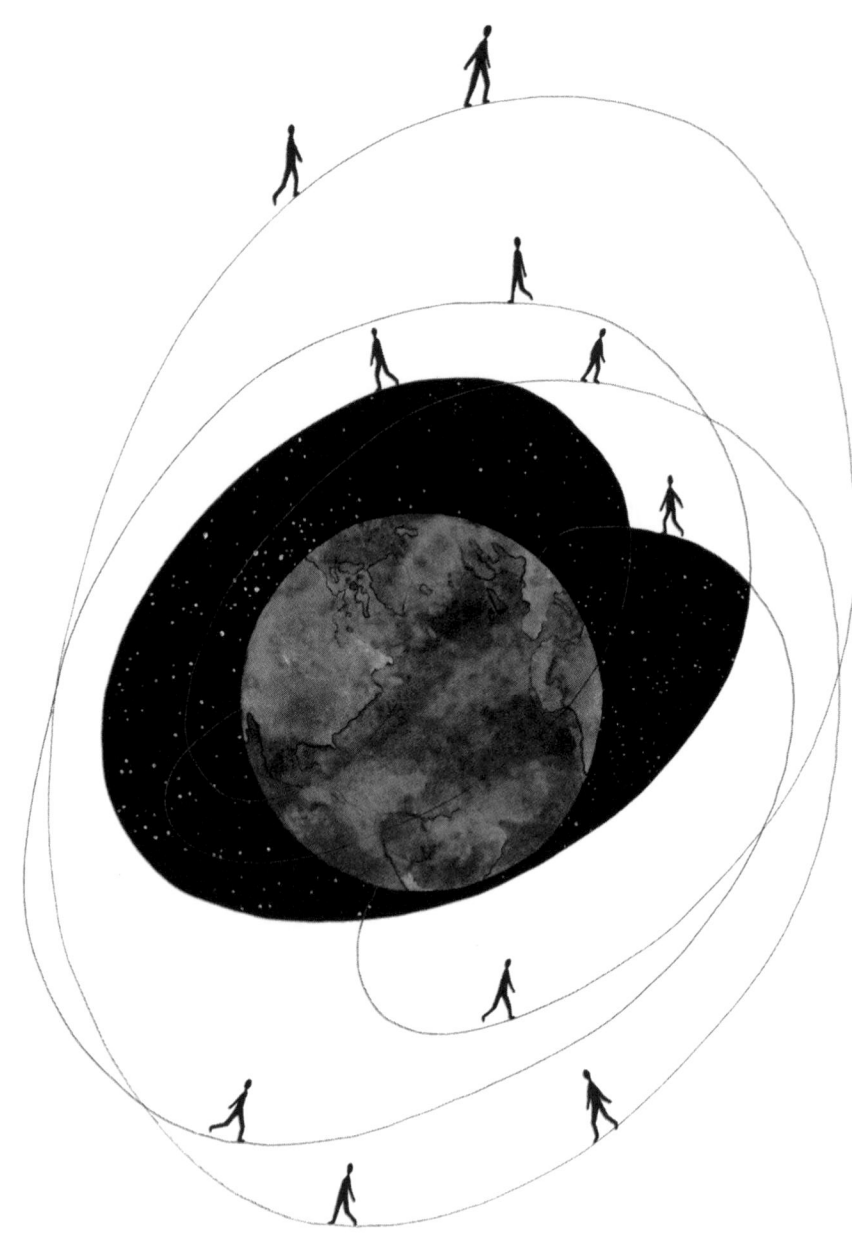

YOU WILL WALK AROUND EARTH
FIVE TIMES

A moderately active person who lives until the grand age of eighty will walk the equivalent of Earth's circumference five times. The equatorial distance around the planet is around 24,900 miles, or 40,000 kilometers, and such an individual could have covered more than 110,000 miles during their lifetime. Such facts seem to alternately expand and condense distance—the distances and corners of a planet that took humans over 85,000 years to inhabit, to settle into.

We have known that Earth is round for thousands of years, with the Greeks figuring it out long before the first circumnavigation of the globe, but it was Isaac Newton who first proposed the idea of it not being perfectly round. Its shape is what is now known as an "oblate spheroid," which describes a sphere with a sort of slight squashing or flattening at its poles; in the case of Earth this is caused by its rotational velocity, which is a nauseating 1,000 or so miles per hour.

If one were being particular, Earth isn't a perfect oblate spheroid either, because its mass isn't evenly distributed—in the more concentrated areas of mass, such as the heaviest of mountain ranges, there is a stronger gravitational force. These anomalies get changed ever so slightly over very long periods of time by the appearance and disappearance of mountains and valleys, the occasional and rude meteor crashing into the surface, the pull of both the Moon and the sun, the fluctuating weight of the oceans and the atmosphere. The entirety of Earth's surface actually attempts to smooth out this irregular distribution itself, to stabilize, a process called "true polar wander."

In order to keep meticulous track of Earth's shape, scientists have now covered its surface with thousands of GPS (Global Positioning System) receivers, which can detect minuscule changes in their height and position. They also fire visible-wavelength lasers at satellites, and have radio telescopes listen to extragalactic radio waves. This all sounds very exciting, and also seems like the least we can do—keep an eye on things, that is.

2,600,000,000 HEARTBEATS

Sometimes it seems strange that we should only get one heart, that we have no choice but to rely on it so heavily, and for so long—that for the most part it keeps time and a rhythm, putting up with our poor choices and our first loves. On average, it beats 4,800 times an hour, which looks like 40 million times a year and 2,600,000,000 times in an average lifetime. Unfathomable though it might seem, the smallest mammal, an Etruscan shrew, has a heart working twenty times faster than yours at nearly 100,000 beats an hour.

A hollow muscle located a little to the left of the center of your chest, roughly the size of a clenched fist, 5 inches long by 3.5 inches wide, and 2.5 inches from back to front, the heart is an intricate system of atria and ventricles and the same decision over and over again: the potential for everything.

The heart's natural pacemaker is located in one of the chambers, the right atrium, and is called the "sinoatrial node"—the electrical impulses that instruct your heart to beat originate here, although other nerves can change the rate of this instruction or alter the strength with which the heart contracts. If they were laid out end to end, the nutrient-carrying blood vessels contained within a human body could wrap around Earth twice, and it only takes one minute for blood to make a round trip from the heart and back again. The sound of a heartbeat, something we sometimes take note of but seldom appreciate, is formed of two parts: the first part is the sound of the tricuspid and mitral valves closing, the second part the aortic and pulmonary valves shutting—these sounds are known as S1 and S2.

It will likely not come as a surprise to you that you can, to some extent, break your heart. Deep emotional pain and distress

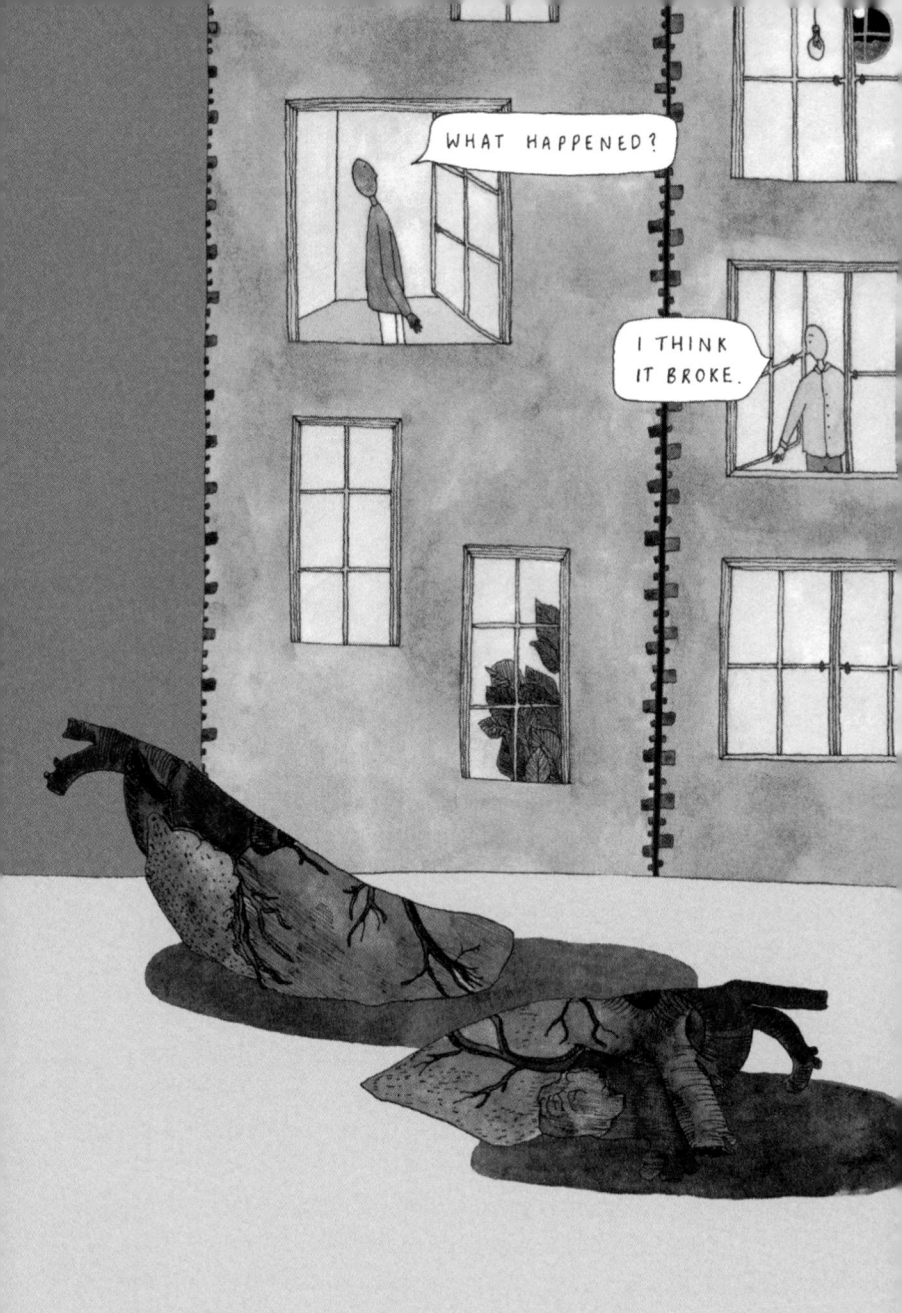

cause the release of the hormone cortisol, which can damage a heart, and brain imaging has shown that the same neurological pathways are lit up by your heart breaking and by the pain you feel when holding on with bare hands to a much-too-hot cup of something.

Please use caution.

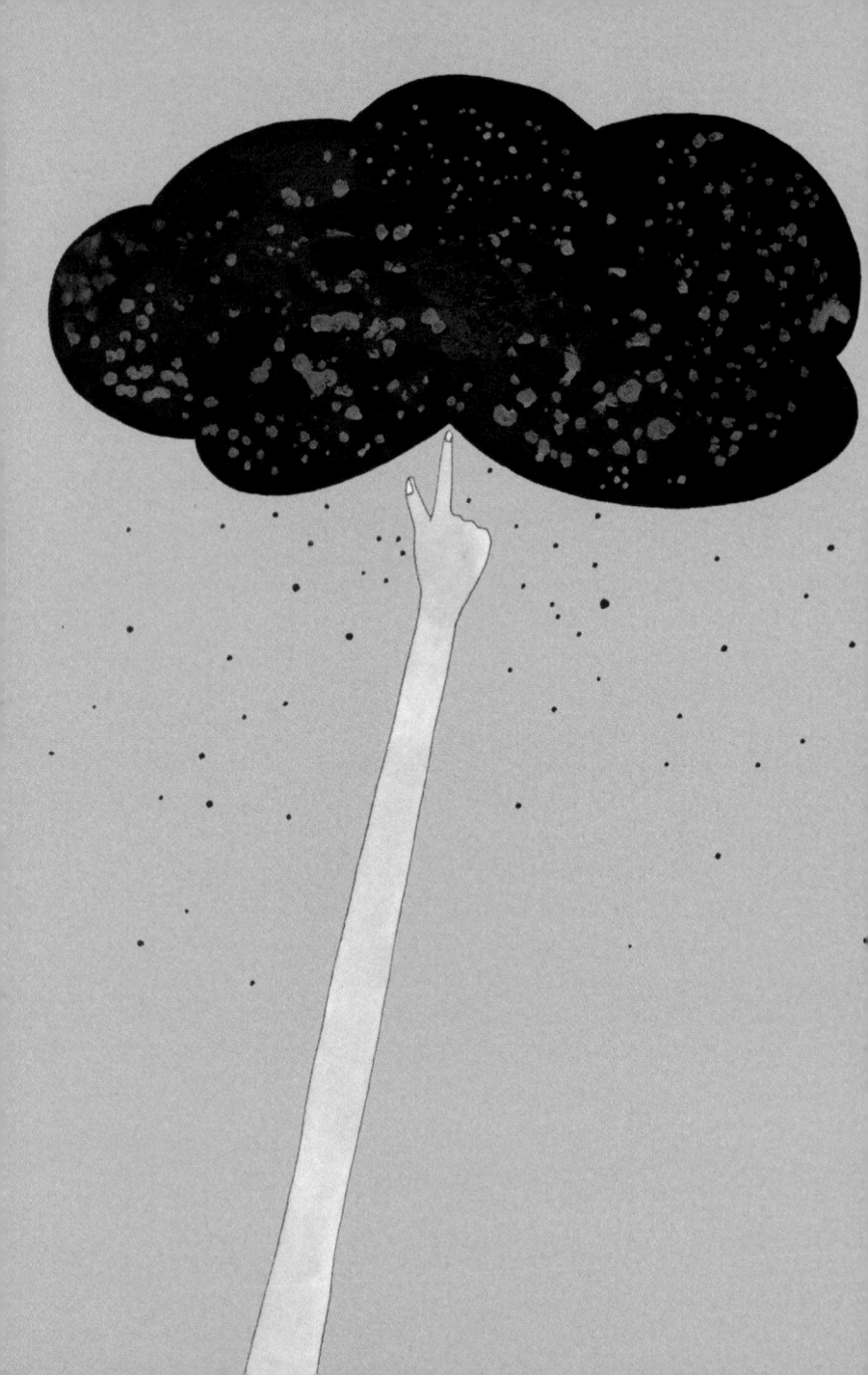

NEVER TOUCHING ANYTHING

Humans are little more than walking patterns of energy, banquets of atoms and their electron clouds (see page 67). It is because of these dancing electrons that objects appear to be solid, tangible; the electron clouds in you and the electron clouds making up everything else are interacting, specifically by not ever really wanting anything to do with each other.

Electrons cannot all occupy the same energy state—some complicated combination of quantum mechanics and the electromagnetic force holding all the electrons together but apart means that at an atomic level, you're never actually touching anything. The indistinguishable goings-on at this impossibly small scale leave a tiny, tiny, atomically small gap in between things—between your hands and a book, between each of the pages of a book, between your feet and the floor. You are not doing anything except holding on to a misleading collection of matter, and it is the same when you cut into or through something—the knife or scissors don't actually come into contact with whatever it is that you're cutting, they are simply pushing atoms out of the way.

The electromagnetic force holding the electrons together is absurdly strong, around 10^{36} times stronger than Earth's gravitational field (in terms of zeros, this is about 1,000,000,000,000,000, 000,000,000,000,000,000,000 times stronger). Within an atom, the electrons stay resolutely away from one another, determined to never touch, and yet are unable to stay away completely; they know even without looking where all of the other electrons are, but they feign indifference, remain persistently aloof.

All we are feeling then, your hand in mine, is our electron clouds and the electromagnetic field; the sensation of touch only the repulsion between things far smaller and far more important than us.

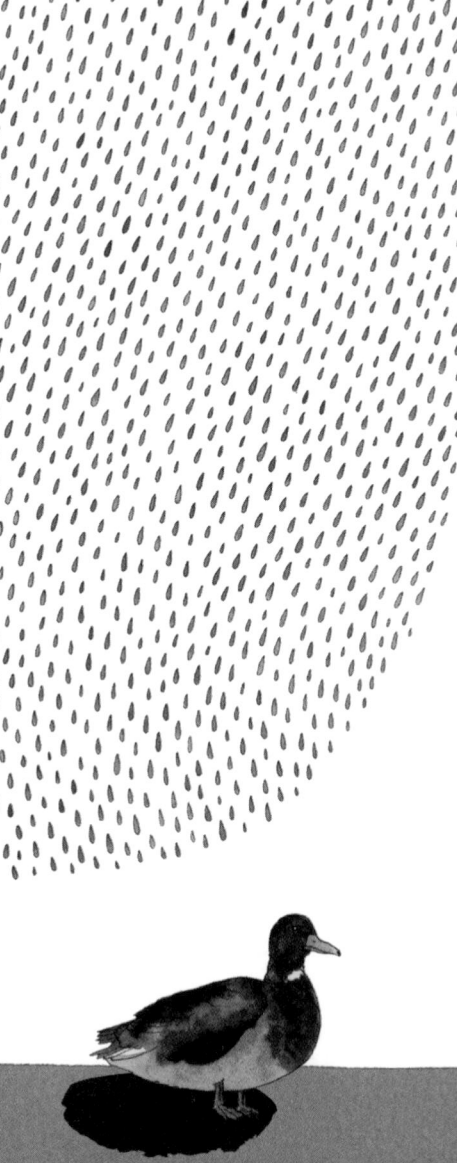

The amount of rain experienced by different places on the planet varies drastically, and the variability of weather seems a universal conversational topic. The highest average rainfall of all is found in a small village called Mawsynram, which is perched high on a ridge in India's Khasi Hills. The driest (and also darkest) place on Earth is Ridge A in Antarctica; the air there is one hundred times drier than it is in the Sahara.

When the water on the surface of Earth is heated by the sun, it evaporates, rising into the air and turning into droplets of water vapor. When this reaches a certain height it will then cool, and condense around particles in the air, becoming visible as clouds (more on this on page 37). Warmer air can hold more water vapor than cooler air, but when air of any temperature can't possibly hold any more moisture, it is saturated, and would in theory have a relative humidity of 100 percent—the term "humidity" refers to the amount of water vapor present in the air.

There are two ways for rain to reach you: the collision coalescence process and the Bergeron process. The first describes how rain forms in clouds that sit well below the freezing temperatures of the upper atmosphere, and involves larger droplets within these warmer clouds colliding with their smaller neighbors and joining together, or coalescing. A million or so droplets later, this amalgamation is then heavy enough to fall as rain to earth without evaporating on the way down.

In the upper, much colder layers of the atmosphere, rain forms differently. Known as the Bergeron process, this is responsible for most of the precipitation that greets us. In the air, pure water doesn't freeze until temperatures reach –40 degrees, and so "supercooled" water droplets will surround the ice crystals within these colder clouds, which grow in size until a snow crystal has been formed. Once heavy enough, this droplet is then able to fall, gaining size as it goes. If the temperature all the way to the ground is freezing, it will fall as snow, but there is often warmer air on the way down and it will therefore melt and fall as rain.

Some of the most well-known and wondrous rainmakers are altostratus clouds (light precipitation), stratus clouds (normally no more than drizzle), nimbostratus (the chief maker of rain), and cumulonimbus (the usual culprit when it comes to heavy downpours and thunderstorms).

Regardless of whether or not you can ever remember their Latin names, and whether or not the cloud is a barely visible wisp or a threatening gray giant, the rain they throw in our direction brings with it some very specific aromas.

In the moments and minutes just before a storm, some people will be able to detect ozone, a form of oxygen whose name comes from the Greek word *ozein* (to smell), which is created by lightning tearing up molecules of nitrogen and oxygen and seems a sort of tipping point, one that you can almost taste. Most of us will only notice it after rain, though, especially after very heavy thunderstorms, and as a clean, almost chlorine-like smell. While we can quite easily detect the scent of ozone, the human nose is far more sensitive to the smell of geosmin, which is a metabolic by-product of bacteria in the soil—an earthy, damp scent that leaves a person feeling as clean as if they had been dragged backwards through a cloud. It is geosmin, along with after-the-rain oils secreted by certain plants, that provides the smell most commonly associated with wet weather. In 1964, two Australian researchers, Isabel Joy Bear and Richard G. Thomas, coined the term "petrichor" when trying to figure out where exactly the smell of rain came from. Petrichor (built from the Greek words *petra*, meaning stone, and *ichor*, which is the blood of the gods in Greek mythology) is defined as the pleasant smell that accompanies rain after a dry spell, or simply as rain falling on dry soil.

EVOLUTION

Evolution is one of the most important advances of modern biological thinking. The term refers to the connectedness of all living organisms, the gradual changes within species and their diversification over millions of years—the process of becoming what we now see when we look in the mirror.

Relying on something called natural selection, evolution as a theory was laid out in the mid-19th century by Charles Darwin in his book *On the Origin of Species*, the contents of which became the foundation of evolutionary biology. Darwin introduced the idea that populations evolve slowly over generations, and although within a couple of decades there was widespread scientific agreement, it wasn't until the 1930s and '40s that his concept of adaptation became quite so concretely central to modern theory. It is now one of the most unifying philosophies for the areas of science studying life.

Evolution is seen when the variation within a species leads to organisms that are over time better suited to their environments: less likely to be susceptible to disease, more likely to survive. These characteristics are more likely to be passed on to offspring, leading to dramatic changes, ones that are smoothed out and rendered imperceptible until looked back at over millions of years.

Scientists estimate that there could be around 1 trillion different species on the planet, with maybe as little as one-thousandth of one percent having been described and studied by humans. But all of this life on Earth, if you go backwards far enough, shares a common ancestor, known as the "last universal common ancestor." This ancestor, whatever it was exactly, would have lived around 4 billion years ago.

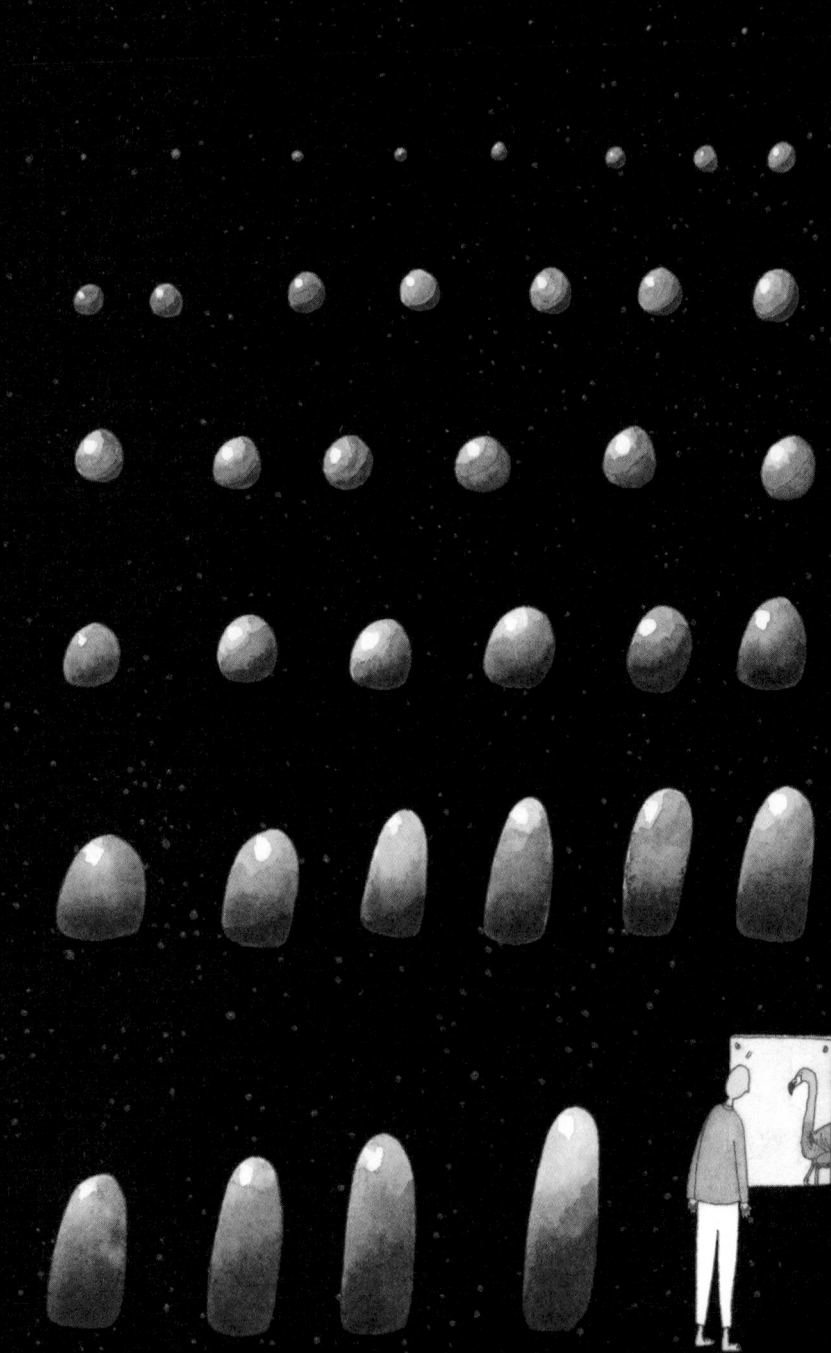

In terms of humans, we weren't alone until relatively recently; at least a dozen other species of early humans have existed, some at the same time as each other, some even at the same time as *Homo sapiens* (you and me). The only ones to remain, our genetic diversity is surprisingly low when compared to other apes—this diversity is measured using "effective population size," which is the number of individuals needed to reproduce the diversity of the entire species. We would need only 15,000 people to re-create the genetic diversity contained within our whole 7 billion or so humans. For comparison, some species of mice have an effective population size of well over half a million.

Whether you know a little about evolution or you're overly familiar with the intricacies of the process, it's both good and beautiful to remember how ancient the lineages of everything are—that we are all just legs and scales, spine and no spine, breathing atop mountains but not underwater, just wings and leftover wisdom teeth.

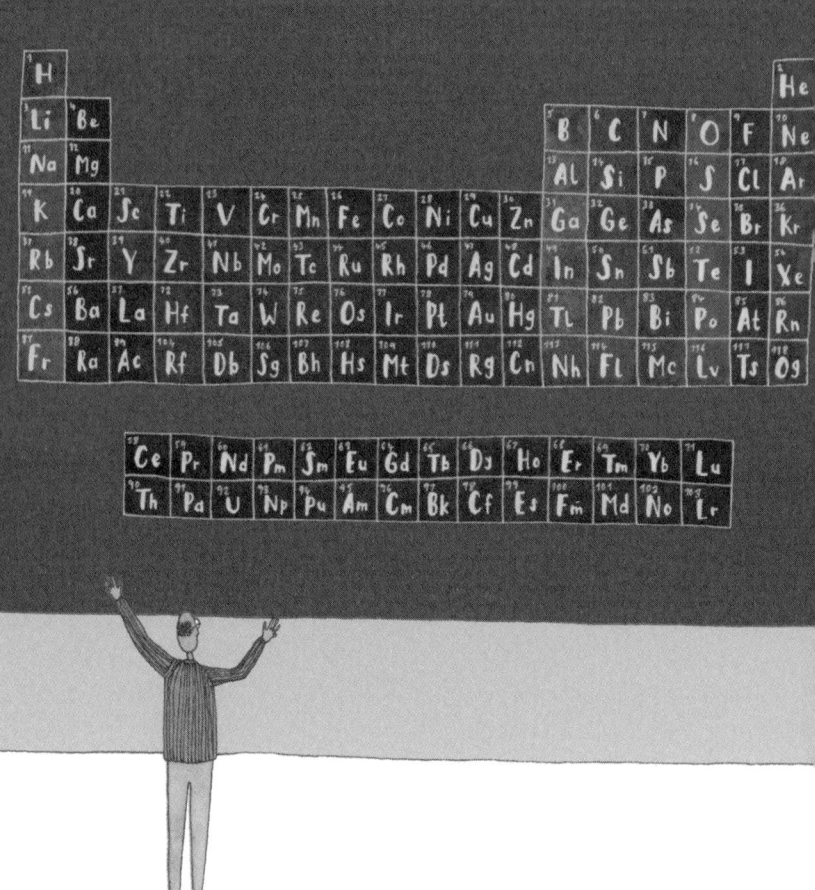

PERIODICALLY

The periodic table is perhaps the most important reference in chemistry, a carefully arranged grid of all the known elements found in the universe. The known elements currently total 118, with about 90 found in nature, and the rest being man-made or created synthetically (elements with an atomic number higher than 92 are man-made). All of them are placed in the table in order of increasing atomic number, which refers to the number of protons in the nucleus. Hydrogen is the first element in the table, as it has a single proton in its nucleus, and the element with the highest number of protons is currently oganesson.

Made up of a series of rows, called periods, and columns, called groups, the periodic table has seven periods and eighteen groups, and the elements in the same group will share similar properties or characteristics. This recurring pattern in their properties is called the "periodic law," and even elements yet to be discovered can have their properties predicted using the table. The most recent four elements added to the table had their names verified in late 2016—traditionally a newly discovered element will be named after a mythological character, a mineral, a place, a property of the element, sometimes a scientist. Oganesson is one of those four, named after Professor Yuri Oganessian, one of the leading researchers in superheavy elements. The other three were nihonium (named after *nihon*, one way to say "Japan" in Japanese) and moscovium and tennessine (both named following collaborations with scientists in Moscow and Tennessee).

A Russian chemist called Dmitri Mendeleev is credited with the creation of the periodic table in 1869 (although versions of it had existed before then, his was the first to gain scientific credibility, and it's still sometimes referred to as Mendeleev's Table).

His original table only contained sixty-three elements, but he correctly predicted the existence and properties of many others. It's astonishingly orderly—all this lining-up helps chemists to understand and predict how an element will interact with others, the types of chemical reactions that are likely or possible for each one.

It's the entirety of our elemental existence, methodically and carefully laid out as letters and tiny numbers, the numbers of invisible things.

THE SMELL OF DYING STARS

Several years ago, astronomers found one of the oldest stars in the known universe; its age is such that it would have formed around the same time as the big bang, some 13.8 billion years ago, and it's located a mere 6,000 light-years away from us. Named SMSS J031300.36-670839.3 (and often shortened to SM0313), it lies in the southern constellation Hydrus and can be seen from here with a largish telescope. SM0313 was formed from the remains of a primordial star at least sixty times the size of our sun.

Until this star was found, scientists thought that primordial stars (the most incomprehensibly ancient ones of all) always formed from large supernova explosions, explosions that throw out, among other things, huge quantities of iron. On closer inspection, SM0313 doesn't contain much of that; it is nearly all hydrogen and helium, which suggests that the explosion of the primordial star that created it was an explosion of much lower energy. SM0313 is believed to be one of the very first stars formed after the universe's initial supernova explosions—a type of star known as a Population II. It is discoveries such as these that allow astronomers to work out the recipes of stars, to map their cosmic fingerprints.

The death of stars is usually something millions of years in the making, and they die because they eventually run out of nuclear fuel. The largest stars die in a far more dramatic fashion than the smaller ones, and the tiniest stars (known as "red dwarfs") burn their nuclear fuel so slowly and steadily that they can live for as long as 100 billion years—far older than the current age of the universe.

The biggest stars, once out of fuel, collapse; the outer layers explode as a supernova and then whatever is left after this explo-

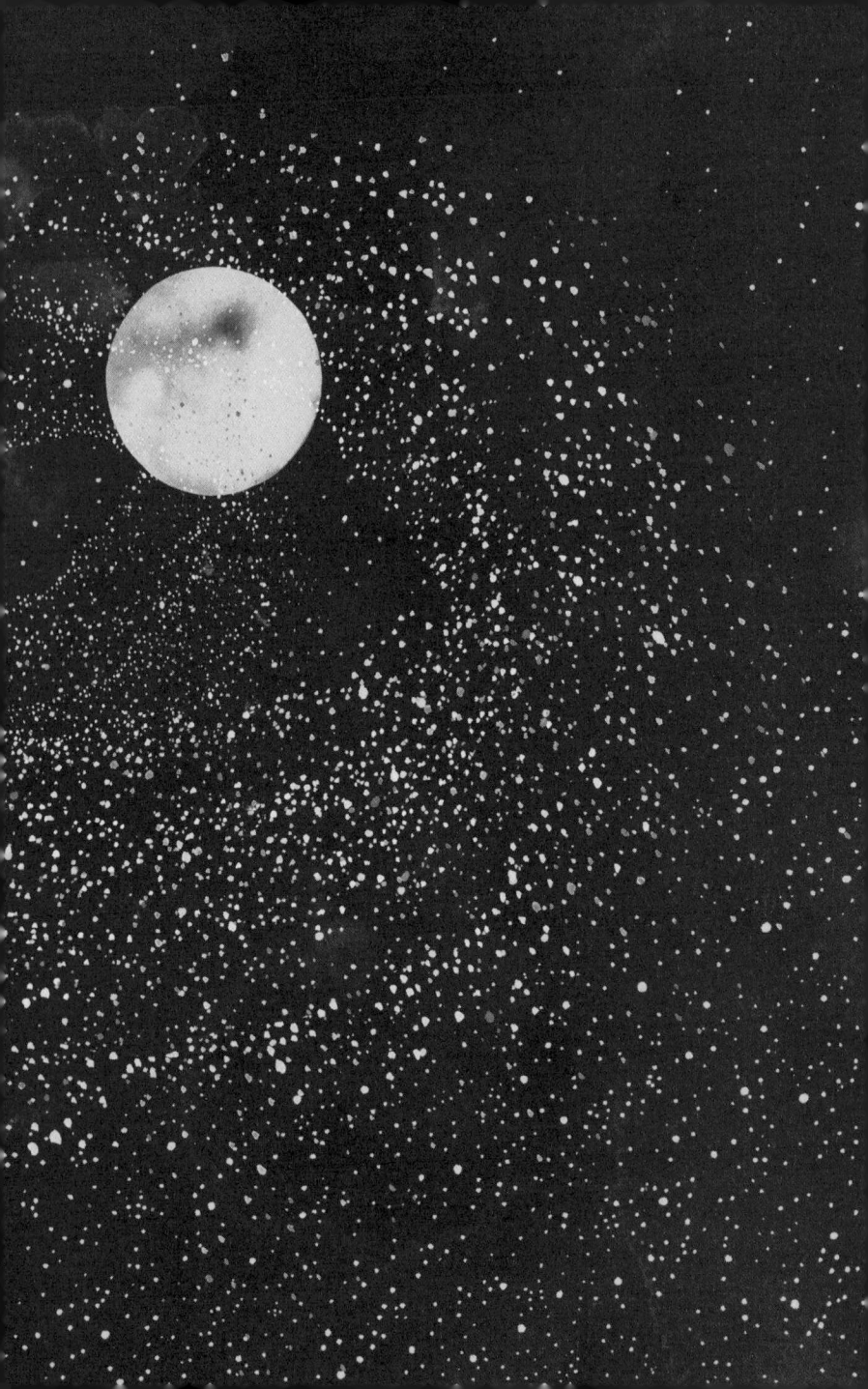

sion becomes a collapsed core, which is then known as a "neutron star." If there is enough mass, a black hole will also form. More-averagely-sized stars swell up to become "red giants" before shedding their outer layers, which often then form "planetary nebulae." The core of these stars will then cool down over billions of years, becoming "white dwarfs."

All of this combustion and spectacular giving-up results in the scattering of compounds known as "polycyclic aromatic hydrocarbons" into all corners of the universe, and it's these compounds that mean that most of space smells like a bizarre feast of hot metal, diesel fumes, and a strangely sweet burning.

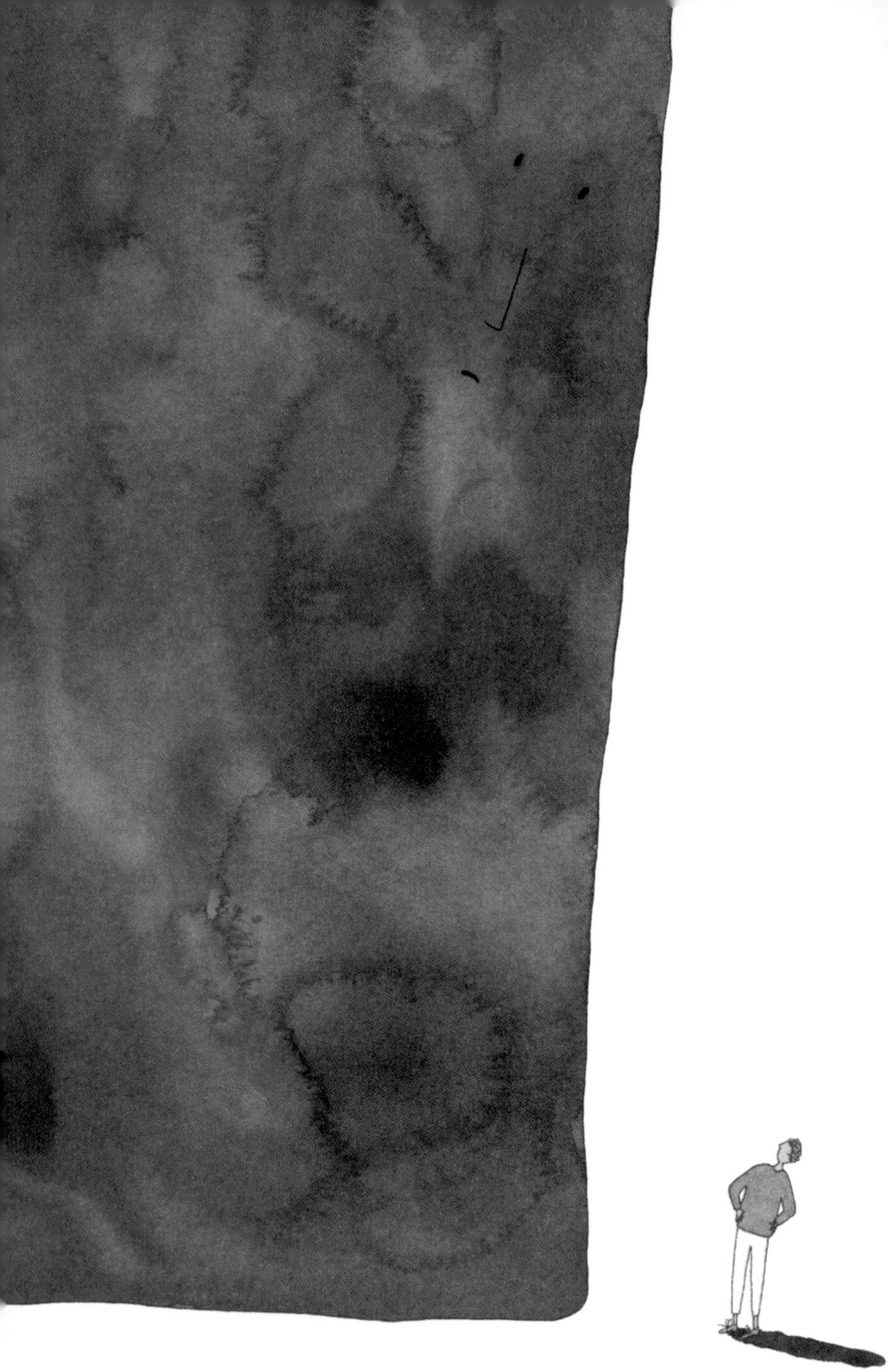

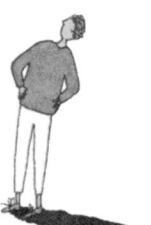

EIGENGRAU

Also known as Eigenlicht, the term "Eigengrau" comes from the German words *eigen* (one's own) and *grau* (gray). It was introduced around 1860 by a German psychologist called Gustav Fechner to describe the disorganized motions of gray seen by the eyes in perfect darkness—the not-quite-black background that is observed in the absence of light. The color of Eigengrau appears slightly lighter than black because for our eyes, contrast is often the most important consideration; the night sky looks black only as the stars provide contrast.

Although the term is now considered dated and is rarely seen in any scientific writing, it's a beautiful example of phosphene—the moving shapes and strange visual sensations of pattern our eyes see in the dark or when we close them. These shapes are thought to be caused by electrical charges produced by the retina of the eye while in its "resting state," a state when retinas are not taking in or processing a whole lot of light or information in the way they do when our eyes are open and the lights are on. Phosphenes are part of a group known as "entoptic phenomena," which refers to the visual effects created by the eye itself.

However, most of the optical phenomena we can look upon are the result of interaction between light and matter. They're often a conversation between the sun or the Moon and the atmosphere—dust, water, clouds, particulates of something or other. These optical phenomena make for an otherworldly list of splendor, a list which includes things such as cloud iridescence (when the colors in clouds resemble those of oil in water); parhelia (also known as "sun dogs," the bright spots that can appear either side of the sun); alpenglow (the reddish, horizontal glow visible on the horizon opposite a low sun, often best seen on mountains); and gegenschein (sunlight scattered by interplanetary dust).

While many of these phenomena are visible with the naked eye, some require precise scientific measurement and observation—one famous example of this is the bending of light from a star near to the sun during a solar eclipse, something that demonstrates the curvature of space itself.

I WOULD LIKE TO PLACE A CALL
TO THE UNIVERSE

It would be fair to say that most people associate outer space with silence, a vast, never-ending, and starlit kind of silent film, a noiseless disco, and this seems reasonable, especially when considering that space is a vacuum (something almost entirely devoid of matter, the word coming from the Latin *vacuus*, which means "vacant" or "void"). But the universe is actually making one hell of a racket, a constant, cosmic pandemonium. In 1964, Arno Penzias and Robert Wilson, astronomers who were using an antenna to keep an eye on their company's satellites, noticed that no matter what they did, there was a persistent and somewhat annoying background noise. What they were accidentally tuning into was hissing cosmic radiation, the microwave background left over from the big bang, the oldest sound of them all, and since then we have been getting to know the precise and peculiar songs of the universe.

Celestial bodies emit radio waves as well as light waves, so we can hear as well as see them. Radio astronomers pick up these electromagnetic vibrations using sensitive antennae and receivers before turning those into sound waves. They're then able to listen to otherworldly music: the solar flares of the sun, accompanied by short bursts of radio energy that to us sound like waves landing on a beach during a thunderstorm; the eerie-sounding storms that rage on the surface of Jupiter; the steady metronome-like thump of pulsars (highly condensed neutron stars containing huge amounts of energy, which causes them to rotate violently, between 1 and 716 times per second); even the icy rings of Saturn, which are truly the sound of emptiness.

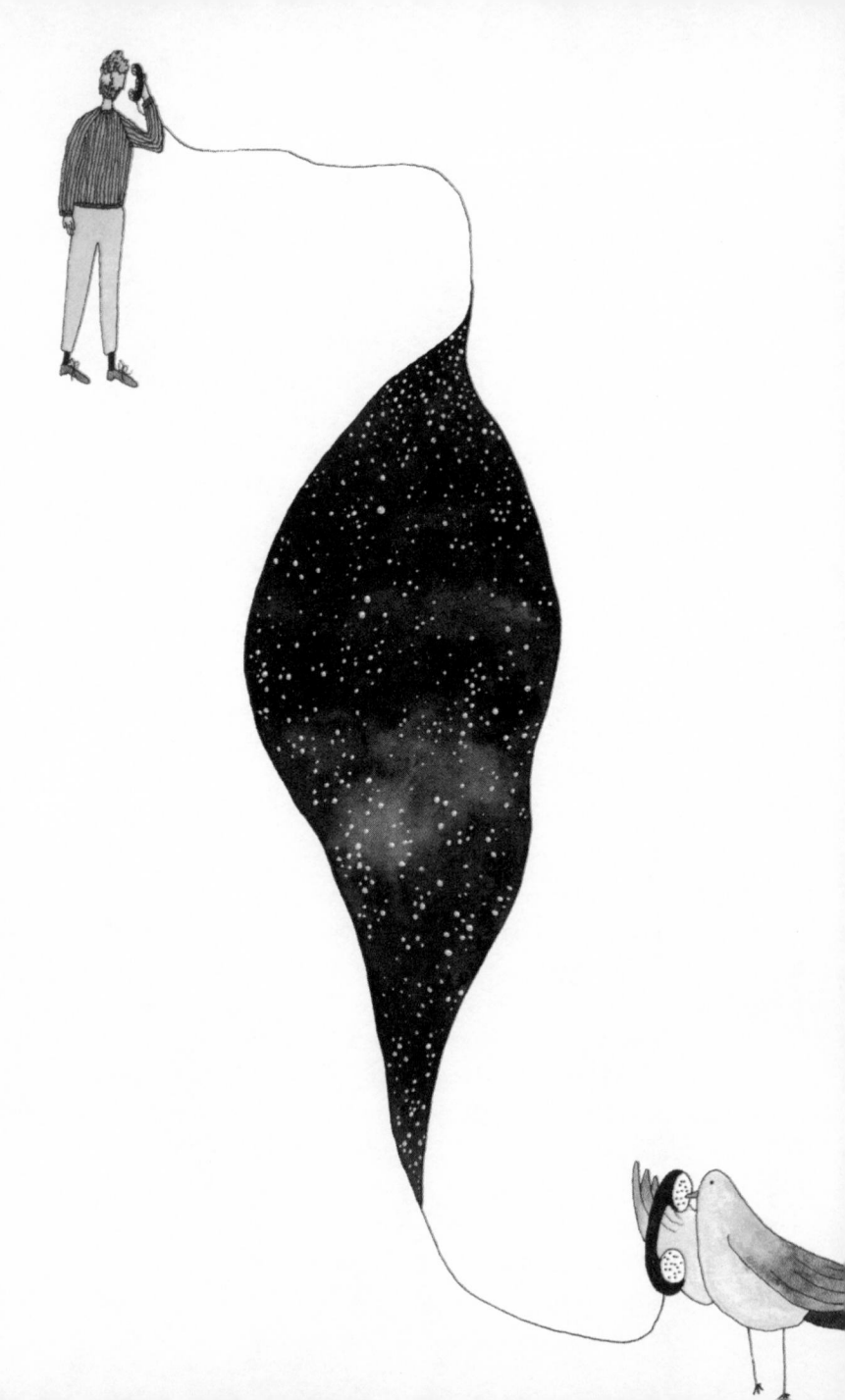

The universe doesn't just make noise; it also has, in a way, a color. If one were to take all of the visible radiation emitted from a huge slice of space, it is possible to then determine how all of that light might collectively be perceived by the human eye. That is exactly what scientist Ivan Baldry and astronomer Karl Glazebrook did, as a somewhat unplanned but interesting side effect of the study they conducted between 1998 and 2003, the "2dF Galaxy Redshift Survey." After looking at the light of more than 200,000 galaxies for five years, they concluded that if the entire sky were then to be blotted out, the universe would be a rather unexciting pale beige, a color since named "cosmic latte"—other suggested options included "univeige," "skyvory," and "astronomer almond," which makes a person feel slightly better about cosmic latte.

Though it may be cosmic latte now, it certainly won't be cosmic latte forever, as the color shifts and changes over time as stars age—young stars are hot and spew wonderfully blue light, while the older, cooler stars emit increasingly red light. In the billions-of-years-ago past, when you wouldn't have been able to see much for all the young, furious stars, the universe would have been a light, cornflower type of blue, and in the next few billion years it will continue to change and, thrillingly, become increasingly beige.

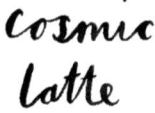

cosmic
latte

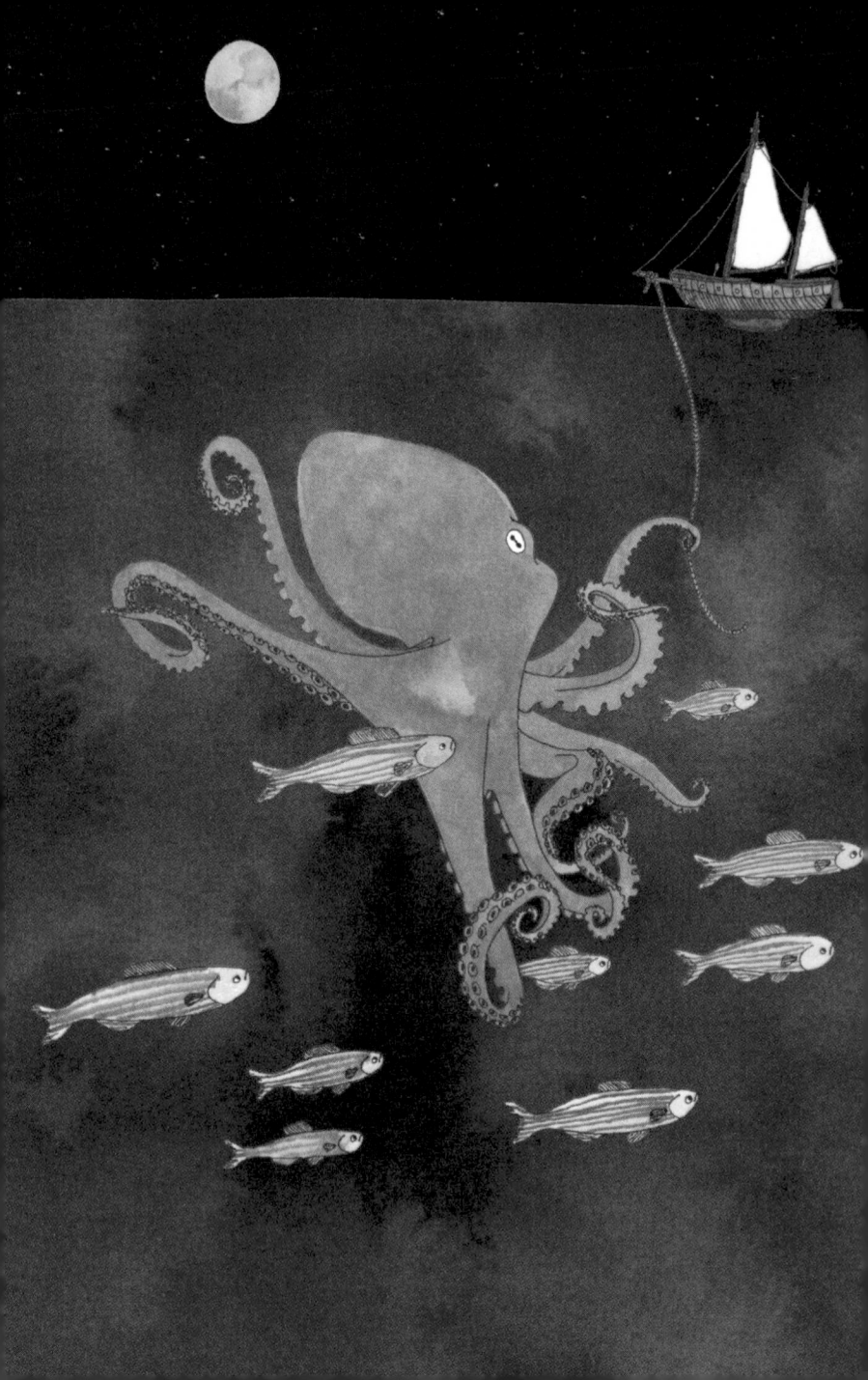

MORE THAN ONE HEART

Having a heart is not always the romantic-sounding business that one might assume, and many animals lack one entirely, having no circulatory system at all. But for others, one is not enough. Cephalopods, like octopus, squid, and cuttlefish, have three hearts each: one systemic heart and two gill hearts ("branchial" hearts), which take blood to the gills on either side of the body. These creatures also have copper in their blood, enough to make them quite literally blue-blooded. And rumor has it that earthworms have five hearts, but depending on your definition, they either have five or none—they have five pseudohearts that wrap around the esophagus.

The heart strangeness by no means ends there. Zebra fish can regrow their own hearts, and even when as much as 20 percent is damaged, the heart can regenerate completely within a couple of months—for a zebra fish, there is no such thing as a broken heart. In the extreme northern temperatures, a species of wood frog, *Rana sylvatica*, can survive freezing conditions because its metabolic activity slows to the point where most of its body water freezes too, and its heart actually stops beating for days, weeks on end. Plenty of animals slow their heartbeat down quite drastically when hibernating, but this frog is the only known example of an animal whose heart stops and restarts with no adverse effects.

There are even a few humans with more than one heart—in cases of extreme heart disease, or cardiomyopathies, doctors can even effectively attach a new, healthy heart to the suffering one, something known as a piggyback heart; the new one can take over most of the workload.

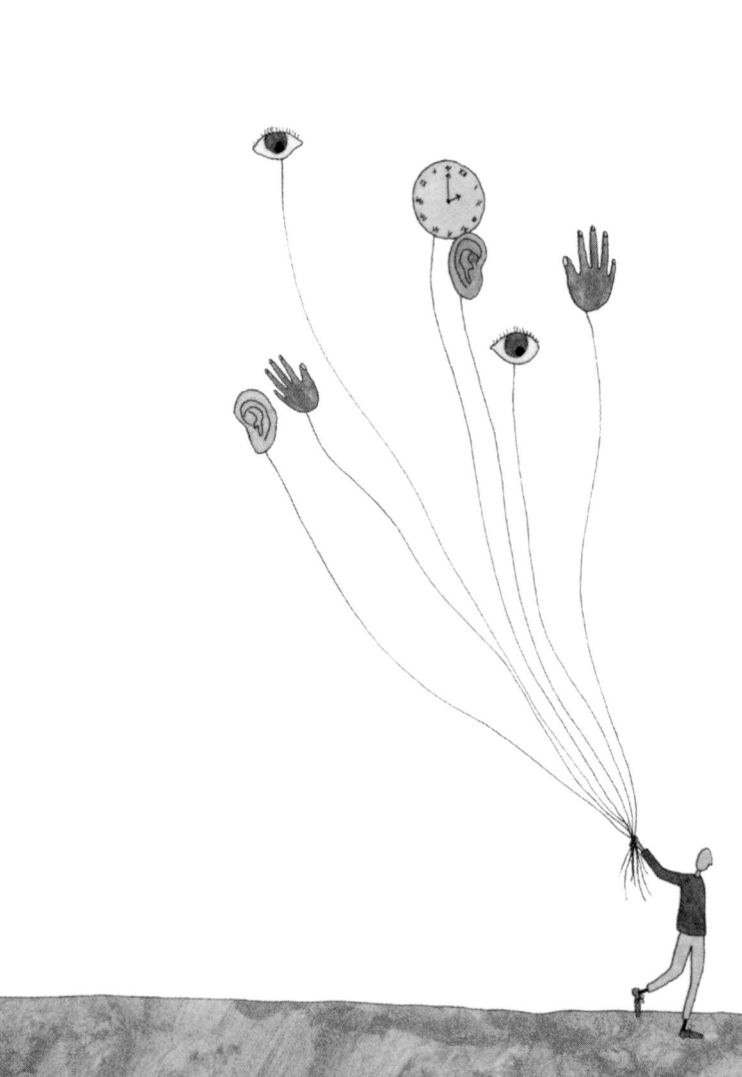

YOU HAVE MORE THAN
FIVE SENSES

During the 4th century, the Greek philosopher Aristotle decided that we possessed five senses, and it was a thought so enduring that many people still speak of and refer to these five senses as if they were the only ones. In the centuries that followed, philosophers were likely missing fundamental aspects of the human experience, as scientists and neurologists now believe that we have many more than five (although their classification is debated and politely argued over, and it is not generally accepted which ones deserve to be listed separately from the traditional five and which ones don't). There could be anywhere between twenty-two and thirty-three senses, all working together so seamlessly that it perhaps isn't all that surprising we blended them.

Aside from the traditional five—sight, hearing, smell, taste, and touch—there are at least a handful of others that are becoming more commonly known. Your sense of balance, for example, is hugely important, and the sense relating to it is called proprioception—the ability to know even with closed eyes where all of your limbs are. Without proprioception, we would always have to walk looking at our feet to know where they were, when they picked up and landed again. There is also thermoception, which enables us to maintain a constant body temperature; making sure we know when an environment is too hot or too cold using thermoreceptors in both the skin's surface and the brain.

The perception of time, chronoception, is arguably also a sense, although nobody knows exactly how it works, and it definitely isn't something found in all life-forms. The fact that it differs between people, and the fact that it's affected and altered by external factors, suggests that we do not have hardwired internal clocks, as it were, but rather we *sense* time, and its passing.

The scientific terms for the five traditional senses make even everyday motions seem quite stately and considered: ophthalmoception (sight), audioception (hearing), gustaoception (taste), olfacoception (smell), and tactioception (touch). Taste, it may be interesting to note, is not the same thing as flavor, which involves both taste and smell—when our 10,000 or so taste buds come into contact with food, they decide between five basic tastes: sweetness, bitterness, sourness, saltiness, and umami. Meanwhile, your nose contains four hundred different types of olfactory receptors and can distinguish between over 1 trillion different odors. Even more wonderful perhaps are fingertips, each of which has two thousand touch receptors—they can detect something as small as three microns high (the width of a human hair is between fifty and one hundred microns).

It is a complicated business, the understanding of our bodies and minds in space and time, and it pays endlessly to notice the slow taste of some things, the faintness of others.

SOUTHERN LIGHTS

The lesser-known but equally astonishing cousin of the aurora borealis, or northern lights, is the aurora australis, or southern lights, and like its northern counterpart, it is the direct result of space weather.

For the most part, a magnetosphere (the area around a planet that is dominated by its magnetic field) protects Earth from things like solar winds and damaging cosmic rays. An aurora occurs when the electrically charged particles from a solar wind make it through. They come from something called a "coronal mass ejection," a large cloud of solar plasma and magnetic field lines, which is thrown out by the sun during particularly strong or lengthy solar flares. These travel over 150 million kilometers before careening into our own magnetic field at speeds of over 6 million kilometers an hour—a geomagnetic storm. Their electrically charged particles and atoms collide with oxygen and nitrogen in the upper reaches of our atmosphere, pushing them up to higher-energy states; the light in an aurora is emitted by those particles when they fall back to their original energy states.

Aurora australis likely receives less attention in part because it generally occurs out over open seas, and to view it from land one would ideally need to be extremely far south—the ends of Australia, Chile, or New Zealand; South Georgia Island or the Falkland Islands; somewhere around the South Pole. Even when seen from land, it can often appear far out into the distance on the horizon. The southern lights also tend to appear red rather than green like the aurora borealis, and our eyes are less sensitive to the color red than they are to green.

If you made it to these tremendous southerly latitudes, you would also be waiting for the clearest of night skies; as auroras

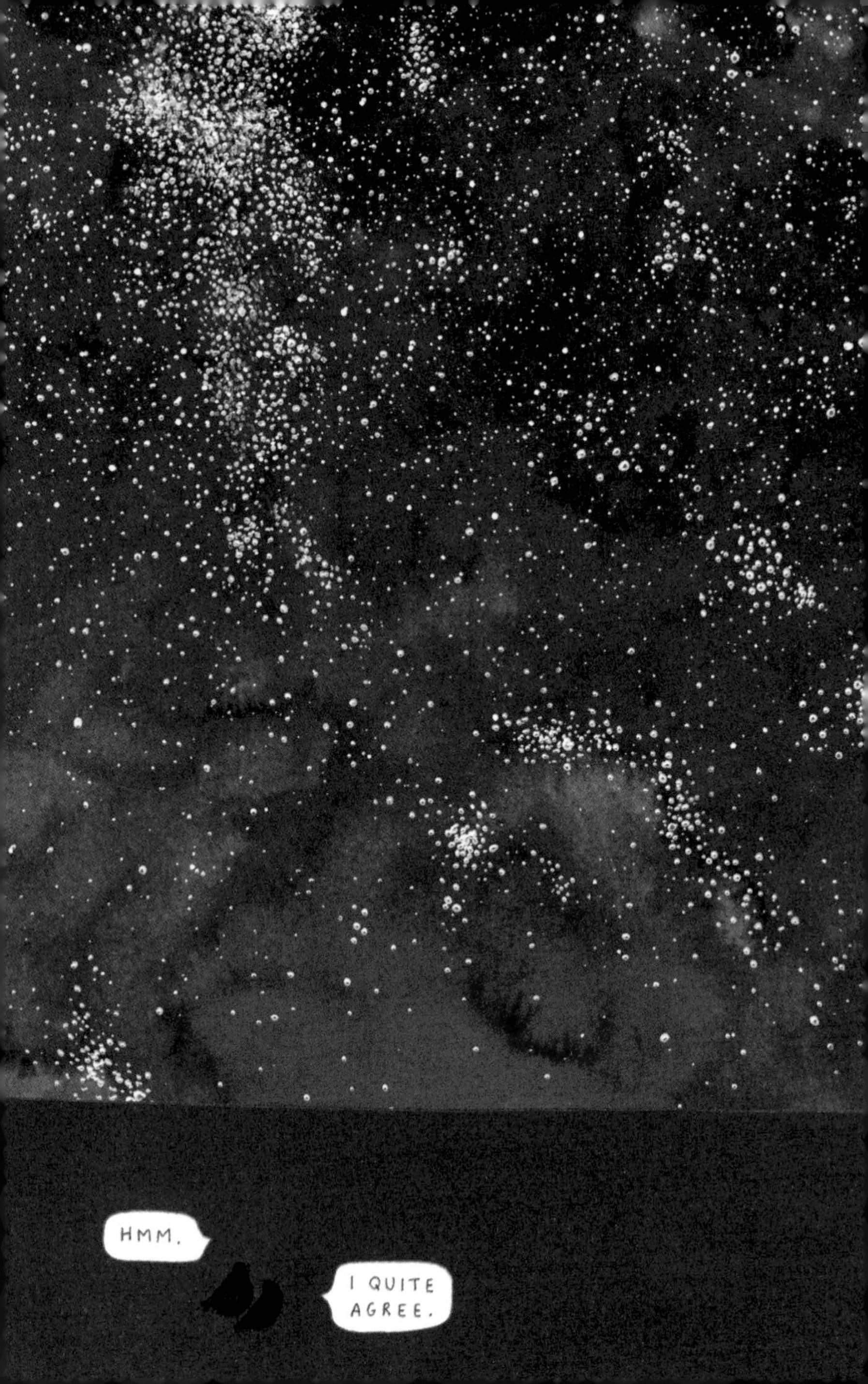

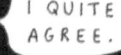

occur in the uppermost part of Earth's atmosphere, you need as little light pollution as possible (that would exclude the Moon), darkness upon darkness upon more darkness. To complicate things further, the sun's commotion moves in eleven-year cycles, and consequently there are periods of higher and lower activity; we're now in the downwards slope of the sun's twenty-fourth cycle (we've only been keeping a record that long), which means we're due for a quieter period of space weather.

We should ensure that we are dressed appropriately.

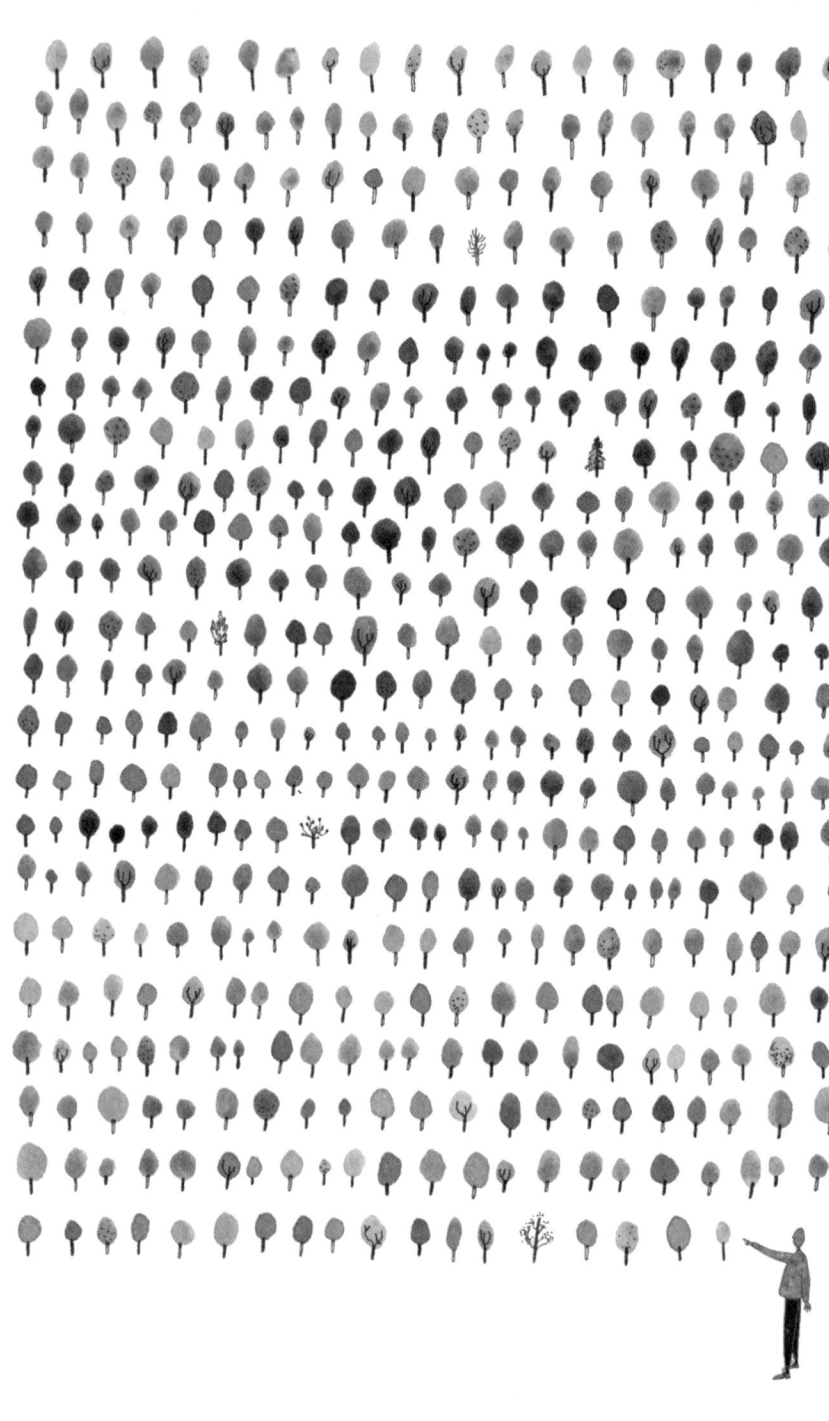

WHAT A DIFFERENCE
JUNE MAKES

There are more than 3 trillion trees on our planet, which works out as around four hundred per person, and there are likely more trees on Earth than there are stars in the Milky Way. Most of the landmass is found in the Northern Hemisphere and so, sensibly, the trees follow suit, with most of the world's temperate forests found in Canada, Siberia, and Scandinavia.

As their lives are ruled by the seasons, the flow of carbon dioxide and oxygen varies quite wildly over the course of a year, and the collective effect that all those trees have on our air is quite astonishing. As they drink in air, they take in carbon dioxide, and, helped by the light of the sun, they store the carbon dioxide, while releasing the oxygen back into the atmosphere.

A single, mature oak tree can have as many as half a million individual leaves, each one covered in tiny structures called "stomata," which act much like lungs. It is the stomata where carbon dioxide gets in and oxygen sneaks out. On a single square millimeter of a leaf, there are between a hundred and a thousand stomata, all equally preoccupied with their contribution to the vast breathing system of Earth.

In winter, without their leaves and therefore largely robbed of their breathing apparatus, trees go relatively quiet, and carbon dioxide loiters in the air—nowhere to go, nothing to do, nobody to take it home. But in June, countless millions of newly grown and green leaves get back to it, and quite literally clean the air of carbon dioxide, something especially noticeable when observing the more northern forests. This cleaning takes place until we reach

November again; then the situation reverses, and the cycle carries on (and on and on).

Although they are not the only monitors of carbon dioxide (oceans and animals also do their part), trees have a noticeably great impact on the planet's health, and they can't keep up with the amount of carbon dioxide that we are pouring into the atmosphere—12,000 years ago, there were around twice as many trees—and consequently, planting them seems a more important pastime than ever.

YOU MAY AS WELL HAVE WINGS

Bones and muscles are similar in many species, often the only difference being the shape of them, or the quantity, or their apparent purpose. It largely depends on what the animal has evolved to do: the equivalent of the human hand can be seen in bats and whales, and the bones are reconfigured slightly to suit a different purpose, while giraffes have the same number of bones in the neck as we do, seven, but we have never needed to use our necks to reach into trees for brunch. When considering birds, our arm is, delightfully, almost exactly their wing.

Move up a little further, and a bird's wishbone is in fact a fused collarbone, serving as a unique attachment for the strong muscles needed to fly. Discounting a handful of dinosaurs, birds are the only animals to have wishbones. However, not all birds have them: flightless birds like penguins lack wishbones, as they simply have no need for such a structure. These fusions of bones are called "ossifications," and they're found in birds more than any other animal. Over millions of years, ossifications have created stronger and more rigid structures that ensure the birds can withstand the demands of flying across oceans, over vast forests, between power lines, and through the dreams of those who want for wings.

Birds also have "pneumatized" bones, which means that their bones contain small, hollow, air-accessible pockets; structures delicate in appearance that both reduce weight and increase mechanical strength—so much so that the total weight of a bird's feathers will often be greater than its skeleton. The type of flight in individual bird species affects the number of "hollow" bones they have; birds that glide or soar for long distances have the most,

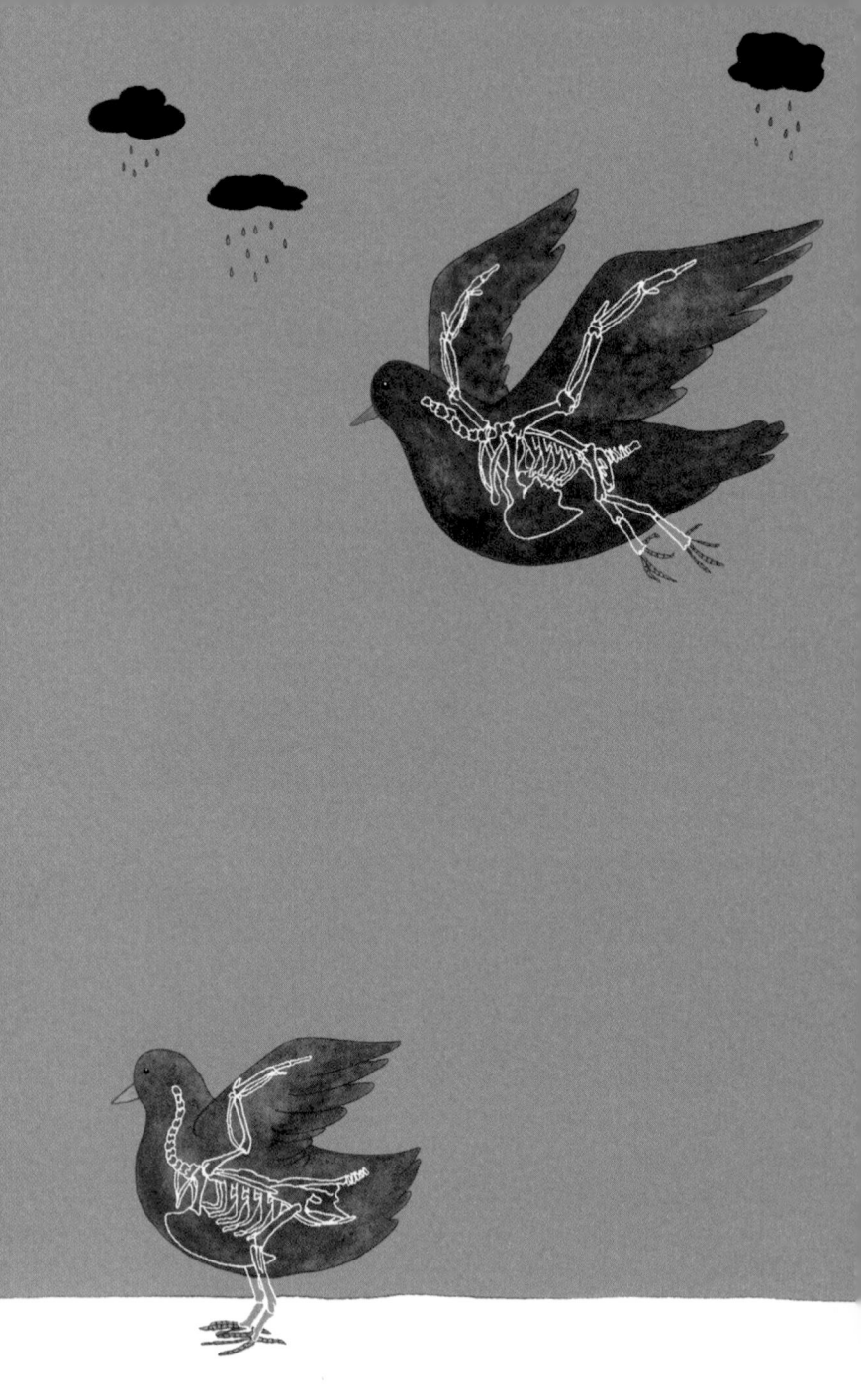

while swimming or running birds, like penguins and ostriches, have none at all.

Apart from these ingenious air spaces within their bones (your bones contain marrow, and terribly essential stem cells), your arm is so similar to the wing of a bird that they are known as "homologous" structures, which means that they are similar enough to demonstrate a common evolutionary ancestor: pieces of a skeleton that are very alike in structure and location in very different species and not necessarily sharing the same purpose. In biology, homology refers to this existence of a shared ancestry between two skeletal structures, organs, or even genes—a leftover togetherness that seems to tie this to that, me to you, you to the birds that circle on warm thermal winds, the atmosphere threaded through their wings, their hearts looking a lot like compasses.

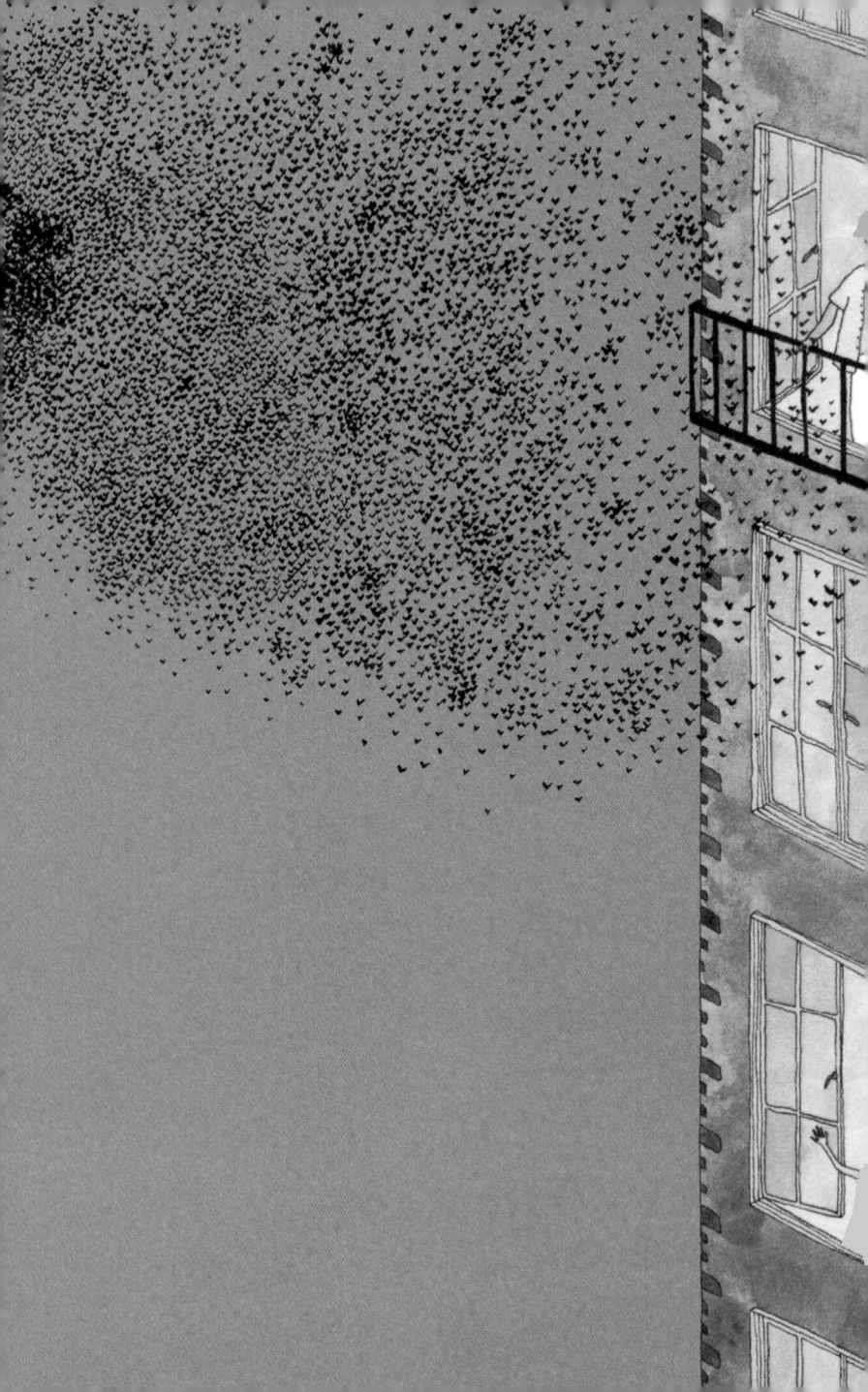

ALL AT ONCE

In nature, two anythings that move or oscillate at about the same rate will, if close enough to each other, gradually begin to move at the same rate or interval. It takes less energy to move together than it does to move apart, or in opposition, a kind of beautiful built-in laziness physicists call "entrainment." Synchronization is one of the most pervading drives in all of the universe, and its reach extends from the tiniest subatomic level to the emptiest corners of space-time. This ultimate tendency towards a spontaneous and bizarre order seems to be the countervailing force to entropy (see page 41), and examples of it are everywhere you look—birds in large flocks, fish swimming in groups to evade predators, tidal rhythms, the marches of electrons, even you.

The natural pacemaker in your heart, called the "sinoatrial node," is composed of 10,000 individual cells, all with their own electrical rhythms that carry signals instructing the heart to beat, but they have to work in unison for you to do anything. Within each of your organs too, the cells are synchronized. There is also the accord required between all of your organs, even though their functions vary wildly. And beyond all of that? Beyond that is synchronization between your body and the world around you in the form of a biological, circadian clock.

It's not only *your* clock that observes synchrony—in 1665, a Dutch physicist called Christiaan Huygens noticed that the pendulums of two clocks in close proximity would end up oscillating together in sympathy, even when they started out with very different intervals. Meanwhile, in Southeast Asia, male fireflies along the banks of rivers flash in unison in their thousands, creating an astonishing spectacle of light that the first Western travelers long thought to be an illusion.

There is even a growing body of research on "interpersonal synchronization," or the ways in which people can synchronize with others and why they appear to do so. Since the 1800s, we've known there is a strong link between respiration and heart rates—an entrainment known as "respiratory sinus arrhythmia"—but there are increasing numbers of studies trying to document it. A Danish study found that putting two strangers in a room to complete a task requiring trust meant that their heartbeats fell into step, while studies in Sweden have noted how the pulses of choir singers speed up and slow down at the same rate, and a US study showed that heartbeats of couples synchronized, even when they were not speaking or touching, and were sitting several feet away from one another.

In many ways, synchrony as a field in science is still at its beginning, but mathematicians and physicists are attempting to understand and pinpoint just how this spontaneous, seemingly effortless order can emerge, unscathed and magnificent, from chaos.

"I want to remember that the sky is so gorgeously large, I feel stranded beneath it."

ANIS MOJGANI

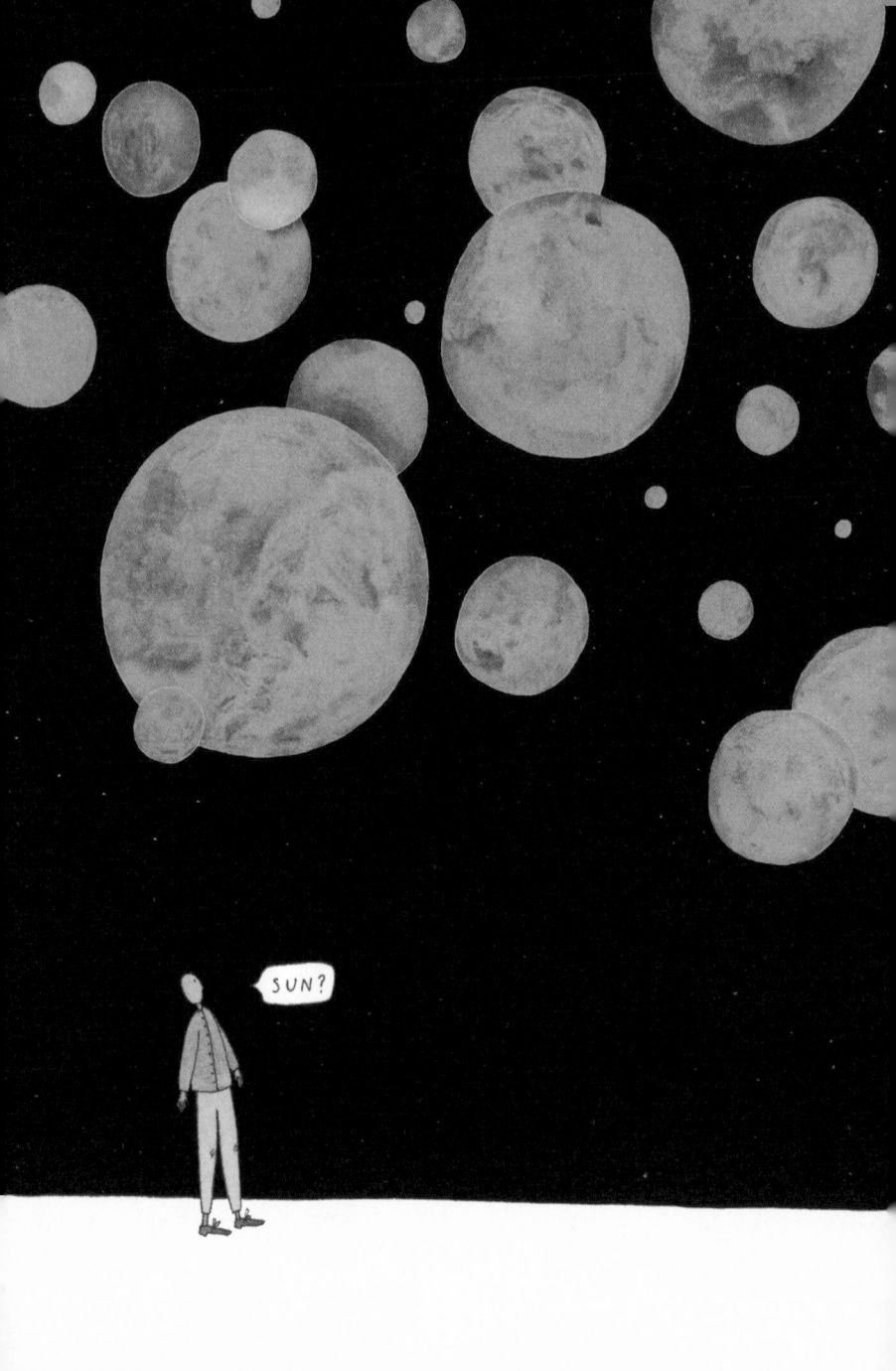

THE SUN IS A TYPICAL STAR

For a long time, it was thought that our sun might not conform to the rules that govern other solar-type stars, but we are, as it turns out, much like the rest.

All stars, however many light-years or dimensions away, have their own magnetic field, and in the case of our sun we now know that this field varies over an eleven-year cycle; throughout this period of time its sunspots, radiation levels, and the amount of material it splutters into space change—the sun alternates between magnetic chaos and extended periods of magnetic calm.

While we can grow to love it, learn to live in the eight minutes it takes for its light to reach us, our sun is only one of countless billions. A typical yellow dwarf star, it was called *helios* by the Greeks and then *sol* by the Romans, hence our arrival at "solar system." In fact, any star can be called a sun if it's accompanied by a planet, or if it's the center of a whole planetary system much like ours, and we are not so unusual in that respect either—scientists know of many "multisun" planets that orbit two, three suns. Thankfully, when in space-time, comparison is no thief of joy.

As previously mentioned (see page 7), everything is orbiting around a center of mass, and the sun is not thinking of us as it circles along with all its coconspirators around the center of the Milky Way galaxy. The time it takes for the sun to complete one such orbit is known as a "galactic" or "cosmic" year, and although the sun may be moving at an inconceivable 828,000 kilometers per hour, we wait about 230 million Earth years for this orbit to happen.

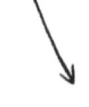

(ABOUT 70%)
DARK ENERGY

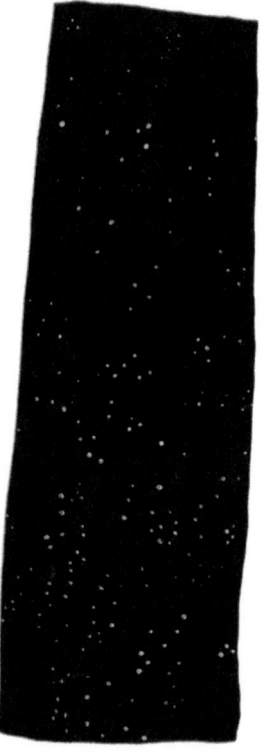

BARYONIC MATTER

(ABOUT 5%)

DARK MATTER

(ABOUT 30%)

ELEMENTARY

Elements are essentially substances that contain only one type of atom. In the context of the universe they are measured in abundance, or how often they occur compared to other elements in an environment. Only a very small portion of the universe is composed from these elements, which is otherwise known as ordinary or "baryonic" matter, while everything else is dark energy and dark matter, both things that have not been directly observed and that have natures yet to be fully understood.

Ninety-eight percent of baryonic matter is made of the elements hydrogen and helium, both things produced in the first minutes of the big bang 13.8 billion years ago. Everything and anything else makes up the other 2 percent, and these other elements are formed when cosmic rays slam into already-present heavy elements like mercury and lead, splitting them into an array of lighter ones such as lithium and beryllium, a process known as "nuclear spallation."

The ten most commonly found elements on Earth are hydrogen, helium, oxygen, carbon, neon, nitrogen, magnesium, silicon, iron, and sulfur, but there are many others that, although relatively fractional in quantity, are very important to life on the planet. Without the element boron, for example, there would be no such thing as a cell wall, not a single plant. It took us until 1928 to realize that copper was essential to humans; it is involved in everything from skin pigmentation to the repairing of connective tissues. Even the most abundant elements are astonishing—ancient stars are the reason why we have any oxygen on Earth, and hydrogen bonds are what give DNA its distinctive twist.

Elements are still being discovered and created synthetically (something that involves crashing different atoms into each other at absurd speeds), but nobody knows quite how much the periodic table (see page 109) can stretch, expand. There are scientists who believe there isn't a limit, but others who think it can only reach so far, that there is a point where atoms simply cannot get any heavier.

Thank goodness for everything that we cannot see and possibly never will.

FIXED STARS ARE NOT FIXED

Fixed stars (in Latin *stellae fixae*) are celestial objects that do not appear to move relative to each other when compared to a foreground of things that do, such as the planets. Ancient astronomers neatly separated the night sky into fixed stars that seemed to rise and fall but stay in certain arrangements, and the planets, which they called "wandering stars." The fixed ones were placed into easily distinguishable constellations, then used for navigation, planning agriculture, keeping to a calendar—the word "constellation" comes from the Late Latin *constellationem*, which means "set with stars."

The total number of stars visible to the naked eye is around nine thousand, and these are all at varyingly huge distances from Earth, contained within our Milky Way but suns of their own systems. In astronomy, stars are referred to usually with numbers, and a great many "star catalogs" have been compiled, constantly changing lists that were being made even by ancient civilizations like the Babylonians.

The problem with the term "fixed stars" is that they do move, all the time, but so slowly that their movements are all but imperceptible on human timescales, these movements so small that they weren't accurately measured until the 19th century. As well as "apparent" motion, which is the illusory movement of unmoving objects in relation to moving ones, stars also have "real" motion; the galaxy containing the star is in rotation, has motion, as well as the star itself within said galaxy.

While absolutely nothing is actually static, motionless, the night sky still looks very similar to how it did thousands of years ago—some of our modern constellations were actually named by the Babylonians; many others still retain their Greek and Roman

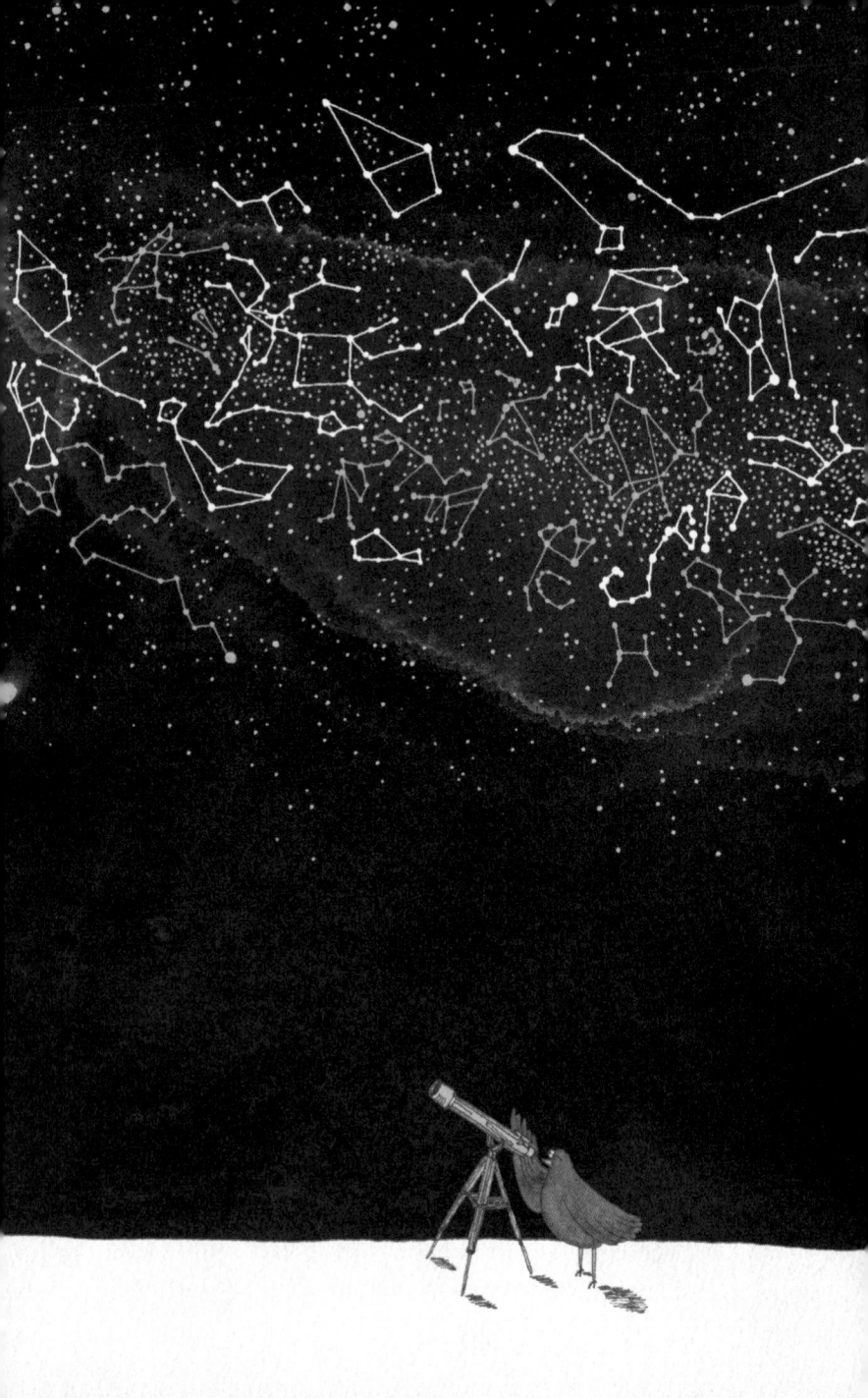

names. The celestial sphere above our heads is filled with eighty-eight official constellations, imaginary patterns or outlines that we can see as orderly pictures because we look up at the night sky in two dimensions, not seeing the depth that in reality puts great distances between all of those seemingly timeless, luminous pinholes.

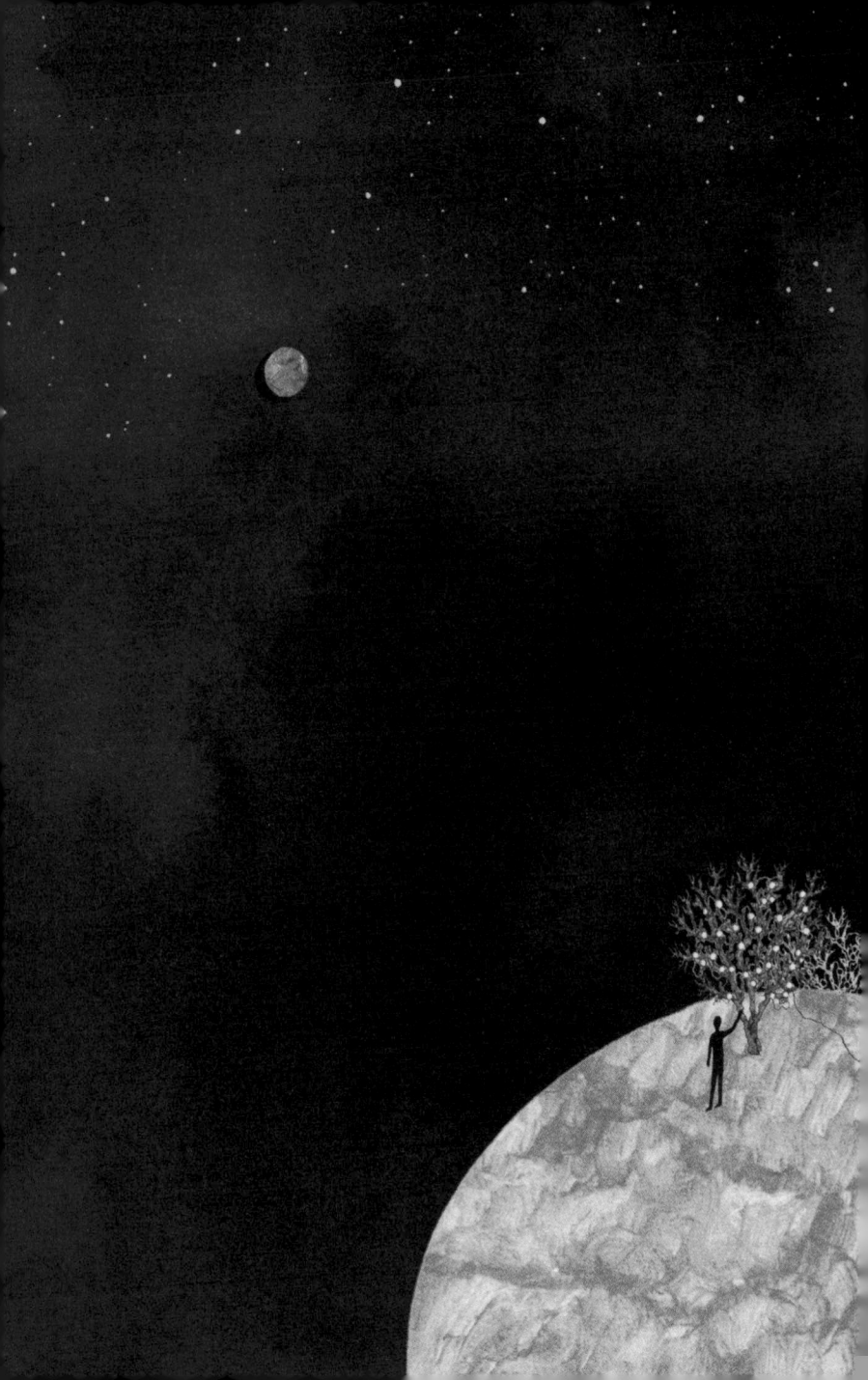

IT WON'T BE TRUE FOREVER

Scientific ideas are far more reliable and far more likely to be true than any other sort of idea; they are based on evidence, peer reviewed, subjected to rigorous and lengthy testing. However, they're not immune to the sands of time and the inevitability that is increased understanding.

Not everything that we currently believe and know to be true will remain correct forever, and in some fields knowledge can become obsolete in a few handfuls of years; information can be disproved or filled out with additional understanding, leading to us knowing the world around us better, with improved accuracy.

Although the path of science leads slowly but surely towards the truth of the matter—of matter in general—we live in a time when knowledge and explanations often change faster than we do ourselves. The branch of science dedicated to this change, scientometrics, is the science of science, and studies how knowledge itself grows and evolves over time.

Facts are disproved at a predictable rate, and their demise follows mathematical curves, something called the "half-life of facts." As with the radioactive half-lives of some unstable atoms, it isn't possible to know exactly *when* a fact will reach its expiration date, but it is possible to observe, when looking collectively at knowledge within a certain sector, a half-life. The half-life of facts is a measurement that shows how long it will take for half of a body of knowledge to become outdated and untrue, although which half cannot be known. In medicine, for instance, the half-life of knowledge is forty-five years, while in mathematics the half-life is much longer—things are rarely ever disproved.

The American writer Isaac Asimov had a wonderful way of putting this phenomenon: "When people thought the earth was flat, they were wrong. When people thought the earth was spherical, they were wrong. But if you think that thinking the earth is spherical is just as wrong as thinking the earth is flat, then your view is wronger than both of them put together."

All this isn't something to fret over; there is as much of a shape and a regularity to both the growth and the dismantling of knowledge as there is to anything else.

ACKNOWLEDGMENTS

I found myself not entirely wanting to list all those who have been instrumental here, because it's commonly known that words are often at their most legless when you're trying (flailing) to explain how magnificent people are.

That being said, I cannot thank the following ones profusely enough: Jennifer Weltz at JVNLA, because while I'm sure everybody says this about their literary agents, you are without question the most wonderful of them all; Meg Leder, my editor-whom-I-find-quite-dazzling, for understanding so quickly what this book was, for being so kind and encouraging; Shannon Kelly, Sabrina Bowers, Elizabeth Yaffe, and truly everybody else at Penguin who stroked the book in some way or another—thank you for allowing me to be pedantic and verge on obsessive about minute details very few individuals are ever likely to notice.

And the others who were unusually important means to this book's end—Catty, for being consistently more excited about things, facilitating my most (and arguably best) ridiculous; Nick, for offering the manuscript-enabling silence of Brazil and all the comfortable silences thereafter; and the countless, nameless strangers who left impressions of varying size and shape on me as I went about the business, the mess, of existing.

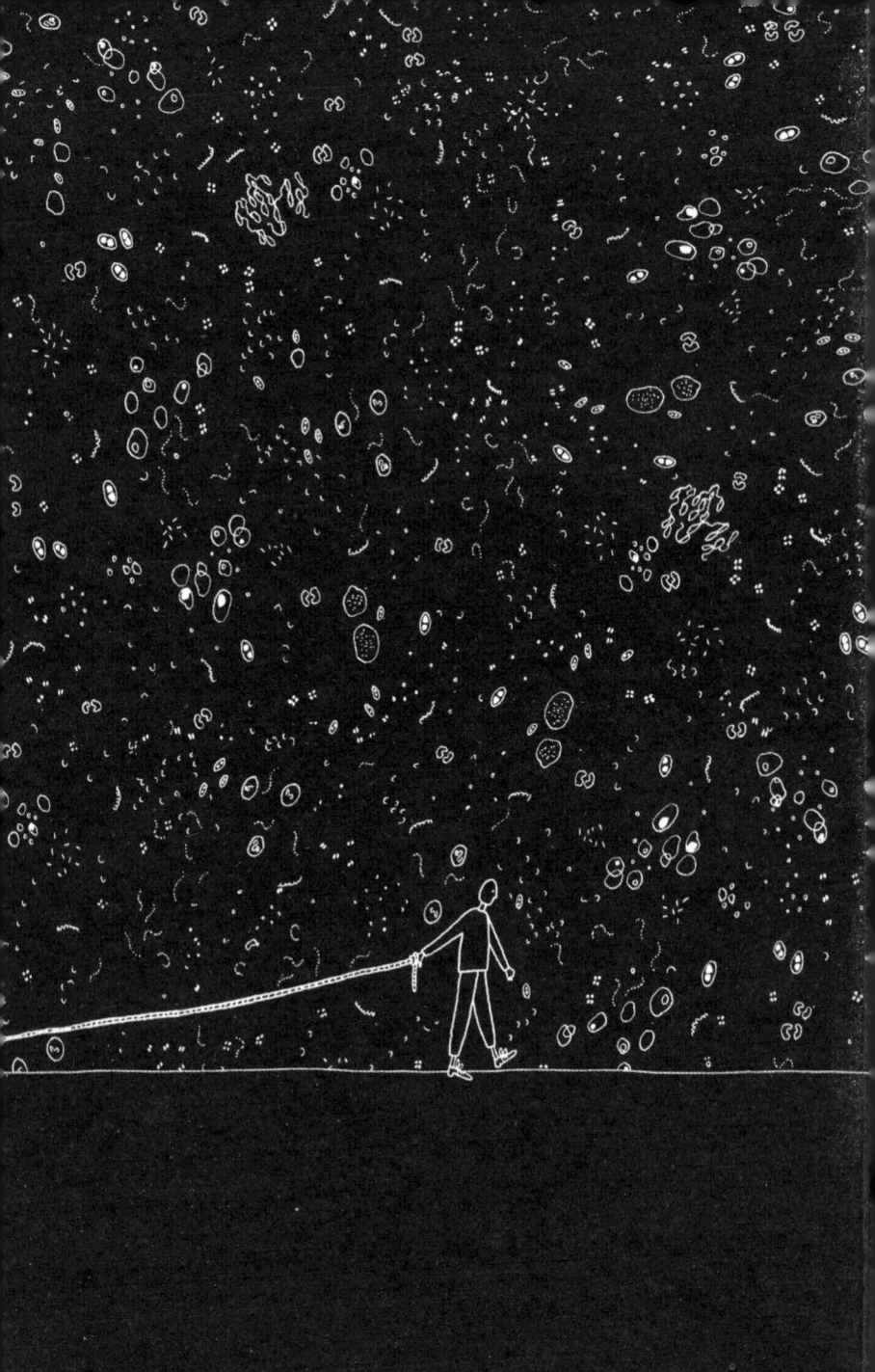